Praise for *Walk*

"It is obvious from the first page that *Walk on Water* will be an amazing book. Ruhlman takes a depressing subject and, while never abandoning hard-nosed journalistic principles, somehow makes it uplifting. He does that by getting inside the minds of Mee and the other medical personnel. Ruhlman deserves the gratitude of readers for allowing them to enter a pediatric surgical unit in such an untrammeled manner. Mee, the other sources who cooperated, and the hospital itself deserve praise, too, for allowing their high-risk world to become known."
—*The Cleveland Plain Dealer*

"Peeling back the layers of the seemingly infallible pediatric surgical unit, [Ruhlman] reveals the human side of a job most of us regard as technical magic and frames a potentially dry topic in a surprisingly intriguing manner."
—*Elle*

"Ruhlman describes with awe the precision, speed and ingenuity required to repair or transplant an infant's tiny heart. His gripping OR scenes capture the life-and-death nature of each surgery and illustrate why only perfection is good enough in this new and rapidly developing speciality . . . most will tear through this engaging and often wrenching account."
—*Publishers Weekly* (starred review)

"Dramatic moments abound as unexpected complications cascade into narrowly averted disasters and last-minute heart transplants, and at these moments the expertise that Ruhlman admires becomes most apparent. A window into an unfamiliar world where excellence is difficult to achieve yet absolutely essential."
—*Kirkus Reviews*

ABOUT THE AUTHOR

Michael Ruhlman, author of *The Soul of a Chef* and *Wooden Boats*, has written for *The New York Times*, the *Los Angeles Times*, and numerous magazines. Winner of the 1999 James Beard Award for Magazine Writing, he lives in Cleveland Heights, Ohio. For more information, go to www.ruhlman.com.

WALK
ON
WATER

The Miracle of Saving Children's Lives

Michael Ruhlman

PENGUIN BOOKS

PENGUIN BOOKS

Published by the Penguin Group
Penguin Group (USA) Inc., 375 Hudson Street, New York, New York 10014, U.S.A.
Penguin Books Ltd, 80 Strand, London WC2R 0RL, England
Penguin Books Australia Ltd, 250 Camberwell Road, Camberwell, Victoria 3124, Australia
Penguin Books Canada Ltd, 10 Alcorn Avenue, Toronto, Ontario, Canada M4V 3B2
Penguin Books India (P) Ltd, 11 Community Centre,
    Panchsheel Park, New Delhi – 110 017, India
Penguin Books (N.Z.) Ltd, Cnr Rosedale and Airborne Roads,
    Albany, Auckland, New Zealand
Penguin Books (South Africa) (Pty) Ltd, 24 Sturdee Avenue,
    Rosebank, Johannesburg 2196, South Africa

Penguin Books Ltd, Registered Offices:
80 Strand, London WC2R 0RL, England

First published in the United States of America by Viking Penguin,
a member of Penguin Putnam Inc. 2003
Published in Penguin Books 2004

10 9 8 7 6 5 4 3 2 1

THE LIBRARY OF CONGRESS HAS CATALOGED THE HARDCOVER EDITION AS FOLLOWS:
Ruhlman, Michael, 1963–
Walk on water : inside an elite pediatric surgical unit / Michael Ruhlman.
p. cm.
Includes bibliographical references.
ISBN 0-670-03201-8 (hc.)
ISBN 0 14 20.0411 1 (pbk.)
1. Ruhlman, Michael, 1963– 2. Pediatric surgeons—United States—Biography.
3. Heart—Surgery 4. Pediatric cardiology. 5. Congenital heart disease in children—
Surgery. I. Title.
RD27.35.R84 A3 2003
617.9'8'092—dc21
[B]   2002033061

Printed in the United States of America
Set in Stempel Garamond
Designed by Nancy Resnick

For Addison and James

A NOTE TO THE READER

In the interest of patient privacy, I've invented names and changed some details for the Scott and the Stone families and the patient referred to as Henry Parker. In three instances I have invented names and changed some details regarding doctors and their centers, noted within the text, and I have abbreviated the names of two families. No other names have been changed.

I am grateful to all the doctors, nurses, and other medical professionals mentioned here for their time and candor. I am particularly grateful to the following parents: Bruce and Karyn Kasnik, Angie and Bart H., Tod and Melanie Deibler, Tim and Kelly Hohman, Tammy Buntura, Edward and Shawna Cundiff, Nina D., Brian and Patricia Mangan, Stephen and Gwen Ferchen, Doug and Cindy Beeman, and Anthony and Tera Anderson.

CONTENTS

1. "Roger, We Got a Problem": An Introduction to the Beautiful, Horrible World of Pediatric Heart Surgery   *1*

2. The Virtuoso   *25*

3. Bad Hearts   *51*

4. Scaramouche   *83*

5. A Beautiful Heart   *99*

6. Serious Business   *157*

7. The Physician's Assistant   *199*

8. "The Physical Genius"   *233*

9. The Good, the Bad, and the Good Enough   *243*

10. The Norwood   *275*

11. What's It All About, Anyway?   *317*

Sources and Acknowledgments   *333*
Resources for Parents   *337*

WALK ON WATER

1. "Roger, We Got a Problem": An Introduction to the Beautiful, Horrible World of Pediatric Heart Surgery

"*Stay away from that.*"

Fackelmann says it to Mac the way he says most things in the O.R.— matter-of-factly but definitively. The two men continue to work within the newborn's open chest, Mac dissecting out heart vessels bound in webs of delicate connective tissue.

Makoto Ando, called Mac, is the chief surgical fellow at the Center for Pediatric and Congenital Heart Diseases at the Cleveland Clinic, in Cleveland, Ohio. Raised and educated in Tokyo, Mac is thirty-five years old, married and the father of a two-year-old daughter, and currently half a year into the fellowship here that will conclude his training as a pediatric heart surgeon. This morning Mac has already accomplished what is by now an almost daily routine for him: the opening of a child's chest. He has drawn a scalpel down the baby's midline, then divided the sternum lengthwise with hand shears.

Mike Fackelmann, a registered nurse and physician's assistant, across from Mac, has fitted the brackets of the stainless-steel chest retractor along each edge of the sternum and cranked a small handle to ratchet the arms open, exposing the chest cavity. Bob Cherpak, the scrub nurse at Mac's right, organizes an arsenal of sterilized steel tools on the setup table and hands Mac instruments as Mac first removes the thymus gland, then opens the pericardium with a bovie—an electric scalpel that cauterizes as it cuts—thus freeing the baby's heart from the blood-bright tissue, a nearly translucent sac. Mac will cut two rectangular patches of this tissue and store them in solution, for use later in the operation. Two stitches are then placed on each side of the opened pericardium and sewn into the patient's chest to hold the pericardium back and present a clear view of the heart.

Roughly the size and shape of a plum, this neonatal heart is pumping at the rate of about 130 beats per minute, normal for a sedated child of this age and weight—forty hours and just over six pounds. The smooth, deep-red muscle on top, the right ventricle, is filled 130 times each minute by the saclike right atrium above it. With the pericardium tied back, Mac begins the work of distinguishing and separating the vessels from one another so the field will be clear and distinct when the chief surgeon, Roger Mee, arrives. Presently, Mac must pull apart or cut with the bovie the mesh of tissue joining the vessels that rise out of the heart, called the great vessels: the aorta and the pulmonary artery.

Fackelmann says it again: "Stay away from that."

Mac, hunched like a cane over the patient, the personification of Japanese silence and humility, says nothing.

The bovie is about the size and shape of a pen. Mac depresses a small rectangular button with his index finger, and the generator issues a high-pitched tone, signaling that juice is running through the chisel-shaped tip. When Mac touches this tip to tissue, the tissue sizzles and pops, and sometimes a wisp of acrid smoke appears. The *blip-blip-blip-blip* of the heart monitor, the *beeeep* of the bovie, and the crackle of moist tissue are the main noises in this bright white O.R.

The baby boy whose heart is open to the room, an apparently normal newborn, is named Connor Kasnik. His eyes, though, are taped closed; his short black hair is matted. His head is cocked to one side, and a ventilation tube has been inserted into his trachea just past the vocal cords; the tube is taped to his mouth. His arms are flat on the table, tiny palms up, a line running into the right radial artery at the wrist. Julie Tome, the anesthesiologist, has also placed a central line in the jugular vein, in Connor's neck, and a peripheral line in a foot vessel. Nothing of this two-day-old baby is visible; everything except his open chest and his beating heart is draped with green cloth, as is the metal cage above his head, where hoses, paddles for shocking, cups, suckers, and assorted tools will rest during the procedure. Julie, stationed behind the cage, has sedated the patient and will determine what goes into his vessels—drugs, saline, albumin, blood—and, until he goes on bypass, what goes into his lungs: oxygen, carbon dioxide, nitrous oxide. She will also draw blood throughout the operation to check the patient's blood gases, which describe how well or how poorly he may be doing.

George Thomas, the perfusionist, is just off the patient's left shoulder, behind the heart-lung machine, which will effectively breathe for Connor and circulate blood through his body. The heart-lung machine allows a surgeon to stop the heart in order to operate inside it (which is why today's procedure is called "open-heart" surgery, abbreviated on the board as MSOH, for "midline sternotomy open heart"), or to reconstruct vessels that would otherwise have copious amounts of blood rushing through them, or to reroute blood entirely via reconstructed synthetic vessels, or to fix valves accustomed to continuous motion. The heart-lung machine is the pediatric heart surgeon's primary tool. George runs through a lengthy checklist, one he completes each time he puts a patient on cardiopulmonary bypass. Lorene Mickunas, known as Lori, tall and slender in sneakers and blue scrubs, is the circulating nurse, assisting all those who are scrubbed, whether to adjust the strength of the bovie, retrieve more 7-0 Prolene sutures for Bob the scrub nurse, or answer the phone—June Graney, a cardiac nurse, will call throughout the procedure so she can update Connor's parents as necessary.

These are the players in this theater this morning. But it's Mike Fackelmann who, though he's not one of the doctors here, seems to command most of the space. Fackelmann stands five feet ten inches tall and has the build and posture of a former athlete who still works out. He wears a shower-cap-style hat that billows slightly over his ears and big, round glasses that enlarge his blue eyes; outside the O.R. he wears small, rectangular frames, but when he's working he needs protection from blood. A pale-green mask, taped over the bridge of his nose to keep his glasses from fogging, covers his nose, mouth, and chin. He wears a standard blue disposable surgical gown, tied in the back, over blue scrubs. Paper booties conceal his immaculate white bucks. Fackelmann, age forty-four, married with two teenage kids, was born and raised in the working-class neighborhood of Parma, southwest of the city, and was a supervisor for a local Ford Motor Company plant, overseeing an eight-cylinder-engine-block assembly line, before a buddy convinced him to become a nurse. At first he pursued anesthesiology, but after spending a year in an ICU and watching surgical nurses in the O.R., he thought, "This is way cooler than putting people to sleep," and changed direction. A certified registered nurse first assistant (CRNFA), casually known as a P.A., or physician's assistant, he's been doing hearts now for nearly fifteen years, and kids' hearts for the last ten. Since 1993, when Dr. Mee arrived at the clinic, Fackelmann has assisted in more than three thousand heart operations. He assists in virtually all of Dr. Mee's cases, and in most of his partner's, when the two aren't operating at the same time. "Roger's my guy," he says.

As Mac continues to dissect out, Fackelmann concentrates on Mac's moves, retracting with a sucker or forceps to help him free the heart and vessels from one another. If this were a normal heart, a thick vessel, the pulmonary artery, would rise out of the top of this right ventricle and divide into two branches carrying depleted blood to each lung. Curving around from behind the pulmonary artery, then arching over it and down behind it again, would be the aorta, the main vessel through which oxygen-rich blood goes to the body. But in Connor's case, these two arteries are side by side, with the slightly larger aorta, a little less than a centimeter wide, rising out of the right ventricle, and the smaller pulmonary artery emerging from the left. In a heart with this defect, called transposition of the great arteries, the main arteries are reversed, so that blue blood circulates continuously through the body and red blood circulates continuously through the lungs. Only a hole that's been opened up in his heart and a fetal vessel called the patent ductus arteriosus have kept Connor from effectively suffocating. The ductus, which connects the two main arteries in utero but then shuts down once a newborn starts breathing room air, has been kept open chemically in this instance. Because lung resistance is high, much of the oxygenated blood courses up the pulmonary artery, shoots through the duct, and enters the aorta, perfusing the body's tissues with oxygen. In Connor's open chest cavity, the ductus is visible as a big, bright bulge between the two arteries.

It's this duct that Mac continues to fuss with. Fackelmann has already said to Lori, "Tell Roger fifteen minutes"—meaning, call Dr. Mee and tell him the team will be ready for him in fifteen minutes. Mac is just trying to make things clear and clean for Dr. Mee, poking around in there with the *beeeep* and sizzle of the bovie. He's noticed that the tissue on the outside of the big duct bulging with all that oxygen-rich blood is bleeding—just a little—and he wants to cauterize it with the bovie.

"Stay away from that."

Fackelmann has said it three times now, as if annoyed—*Cut it out, man, stay away from that*—definitively.

Mac hears the admonition and returns to the task of clearing and freeing. But the little leak of blood along the bulging duct persists, and finally he decides to take care of it. He gives the spot just one more little *zap*—but he pushes too hard, burning a hole through the ductus. Blood immediately fills the field.

As always, Fackelmann speaks matter-of-factly and emphatically, but now he raises his voice to a pitch that signals Julie, the anesthesiologist, and Lori, the circulating nurse, to pay attention. *"We're in the duct,"* he says, getting a sucker in there to pick up the blood that has turned the chest cavity into a bright lake. And then, fixing his eyes on the monitor as the baby's pressures begin a free fall, he says, "Lori, get Roger in here."

Suddenly, several things have to happen in response to the fact that Connor is now losing a lot of blood fast, and the amount of oxygen being delivered to his organs is plummeting. Julie, most critically, has to hang blood, crank the O_2, get pressers into the baby, and clamp his vessels down to maintain blood pressure. Lori's got to find Roger, who as it happens is in his office and, great good fortune, already dressed for the O.R. Mac has to remain calm and keep the bleeding down until Roger can get in here, cannulate— that is, insert the tubes for the bypass machine—and get the kid on pump; once he's on pump, he'll be safe. A visitor arriving at this moment might not even realize that anything is wrong: the room is quiet; no one's screaming at anyone, no one's moving quickly; there's no sign of panic—except that Mac says "Fuck." And then "Fuck." It sounds strange coming from him, again and again, because he seldom utters a word in the O.R. "Fuck." Scrub nurse Bob Cherpak, a Gulf War veteran, is silent, but his every cell is stressed. He knows what he's seeing: a kid dying right in front of his eyes.

"Fuck," Mac says again. He tries to cover the hole with his finger, but there are two problems with this strategy: first, they've stopped giving the boy prostaglandin, the stuff that keeps the duct open, so it's already started to disintegrate and now has the consistency of tofu; and second, the amount of pressure required to stop the blood from spurting out will clamp down the duct entirely, cutting off oxygenated blood to the baby's aorta and therefore to his brain and heart.

Fackelmann keeps looking up at the door, wondering what's taking Roger so long. He'd better get in here fast, he's thinking.

Julie's watching pressures, which have fallen into the 30s. The blood's oxygen saturation is dropping, too. This means there's blood in the arteries, but it's not moving very quickly, and what blood there is doesn't have enough oxygen in it to maintain the patient's tissues. Without oxygen, the brain can go undamaged for several minutes; the big worry here is the heart. The heart needs tons of oxygen because it's a perpetually beating muscle and has a voracious appetite. A web of arteries descending from two main coronary arteries feeds it. The main coronary arteries are in turn fed by the aorta. With decreased pressures and a hole in the duct, the heart's not getting enough oxygen. Its response to this is to beat faster, which only intensifies its need for oxygen that isn't there. This will escalate, the heart beating faster and faster, till it runs out of energy and arrests. And that's what's happening now, to this baby. His heart is working, and his oxygen saturations are dropping rapidly, and his blood pressures are low. When his heart stops, he will die—and since they'll still be many minutes away from getting him on pump, and unable to fix the hole in the duct, the main channel for oxygenated blood at this point, there won't be anything anyone can do about it.

Marc Harrison is thirty-six, wears a short, dark beard, has dark eyes, and on Thursday, December 7, the day before Connor Kasnik's surgery, is dressed in khakis and a tie. A white badge clipped to his shirt pocket identifies him as a staff doctor in the Cleveland Clinic's Pediatric Intensive Care Unit. He has just stepped out of room 10, now occupied by the hours-old Kasnik baby. Connor was transported here in an isolette—an incubator-like tank—from Fairview Hospital, a Cleveland Clinic affiliate, where transposition of the great arteries, or TGA, was diagnosed and treatment with prostaglandin, the hormone that keeps the duct open, begun. Connor is critically ill, and the situation is considered urgent.

Marc loves this part of his job, when he's like a general marshaling his troops—in his case, a small resuscitation team, residents, nurses, technicians, and respiratory therapists. Everyone has his or her job, and the work is calm, so steady as to seem slow, and all but silent. "When it's done well," Marc says, "the initial stabilization of a critically ill child is a beautiful thing, really elegant."

Even before the traditional ABC evaluation (airway, breathing, circulation), Marc makes a visual assessment of Connor as the nurses hook him up to a monitor, then looks at his heart rate and blood oxygenation and watches for any signs of dangerous distress. He'll want to get him intubated and gain vascular access, insert a line into the infant so he can give him fluids. While the nurses check vital signs, the respiratory therapists set up the ventilator and ready the tools for ventilation. This morning, as a snowstorm whips across the room's window, the procedure is routine.

When it's done and the patient is stable, Marc steps outside the room and sees a man with dark hair and a gray complexion walking the PICU corridor, which has glass enclosures on either side. The man looks right and left and right. Marc suspects this is the father. He looks more than just lost, he looks like he's in an awful dream—confused, in danger, vulnerable, eyes red from sleeplessness or tears—but he is also purposeful: he is searching for his new son. Marc steps toward the lost man and asks, "Are you Connor's father?" The man nods, registering a moment of relief—*I'm in the right place*—and enters room 10, one of fourteen in the PICU. When he sees his son, he turns to Marc.

"Hi, I'm Dr. Harrison," Marc says. He shakes the man's hand, identifies himself as the attending physician, and explains what is happening: Connor has just been intubated—that is, has had a tube inserted down his bronchus so he can be ventilated—and is now hooked up to a monitor that will display blood pressures, heart rate, oxygen saturations, and temperature. Marc tells Mr. Kasnik that everything is OK.

A moment passes as Bruce Kasnik takes in this information and the sight of his son. He has driven through traffic-clogged, snowy streets, parked in an illegal spot near the ambulance bay in the hope of seeing Connor as he went in (too late), made his way into the massive clinic complex, and found his baby boy; now he's being addressed by a doctor who seems to be in control of the situation. He can draw a breath, pause, not do anything for a moment.

Marc asks, "How are you doing?"

Bruce Kasnik offers a fragile smile and answers, "Not good."

Marc knows that conversation can have a calming effect on a rattled parent, and he also wants to glean any information he can about the new arrival. In the PICU, doctors don't treat only the patient, they treat the whole family; families come in an astonishing variety and often bring with them troubles beyond those of a sick child. Nothing appears to be worrisome here: Marc sees an appropriately upset, likely exhausted, but coherent parent in clean, casual clothes.

"Is this your first child?" Marc asks.

"No, second."

"Was it a normal delivery?"

"Yes. It was a normal *pregnancy*."

This halts Marc for a moment. "So this was a surprise."

Bruce says, "This was a big surprise."

· · ·

It all seemed to be going so beautifully for Karyn and Bruce Kasnik. Four years ago their son Kyle was born a month premature, and they spent days in a neonatal ICU with him, waiting to see if everything was OK. This time Karyn carried to term. Her OB-GYN never did an ultrasound because, this doctor explained, there were no indications that one was needed. A nervous, expect-disaster-type mom, Karyn signed up for an ultrasound at Cuyahoga Community College, performed by medical students and an instructor; the hour-plus exam showed a normally beating, four-chambered heart, four working valves, two great vessels. The pregnancy proceeded without incident, and the birth itself went smoothly. The couple's family organized a pizza party that evening in the room at Lakewood Hospital. When the last congratulations and hugs had been given, the last relatives had gone, and Bruce was back home with Kyle, Karyn felt exhausted but, remembering Kyle's birth and those days in the ICU, serene. She now thought gratefully, "*This* is how it's *supposed* to be." At around eleven P.M. she gave her baby boy to a nurse so she could close her eyes for a nap. A half hour later the nurse woke her and told her not to be alarmed. They were sending Connor to another hospital right away—they didn't like the blue that had begun to show around his lips and in his fingernail beds. They wanted to make sure nothing was wrong, and they didn't have the diagnostic tools here to assess the situation. Connor could go either to Rainbow Babies' and Children's Hospital in Cleveland or to Fairview Hospital, which was closer. Karyn still isn't sure why, but even though her first child had been born at Rainbow, she chose Fairview, a Cleveland Clinic affiliate. She then called Bruce at home and told him the news. Within minutes Bruce had his sleeping four-year-old and the dog in the car; minutes after that, he had dropped them at his in-laws' and was on his way through winter snow to Fairview. The bad dream, as he would call it, had begun. At 2:30 A.M., a cardiologist diagnosed transposition of the great arteries with intact ventricular septum, a life-threatening condition that would require open-heart surgery. Connor was stable for now, the cardiologist said, but he would have to be transported to the Cleveland Clinic Children's Hospital in the morning.

. . .

Bruce is carrying a book when he walks into the PICU some eighteen hours after the joyful birth. Wedged in the book is a piece of pink paper with a diagram on it, drawn for him by the cardiologist who diagnosed the defect in Connor's heart. Bruce asks to go over it with Marc Harrison and Geoffrey Lane, the cardiologist who got the call from Fairview and has just arrived at the room. Geoff introduces himself, then he and Marc describe TGA to Bruce again: depleted blood enters the right side of the heart from the body, but instead of being pumped through the pulmonary artery to the lungs, as it should be, it gets pumped back out to the body through the aorta; oxygen-rich blood enters the heart from the lungs, but instead of being pumped to the body, it's pumped back into the lungs. In other words, there are in effect two separate circuits of blood where there should be just one.

Bruce nods but is apologetic: "I'm a little blurry. It's been a long night."

Geoff Lane, a thirty-nine-year-old Australian with an oval face, short gray hair, and a calm manner, explains, nodding continually, that they will first do an echo to see for themselves what Connor's heart looks like, and then they will likely have to do a procedure, here in this room, called a balloon atrial septostomy, to open up a hole between the two atriums that will let oxygenated blood into the right side of the heart, thereby increasing oxygen saturation in the blood going out to the body.

Geoff says to Bruce, "The fortunate thing is that for this kind of problem, you're at probably the best place in the world."

No one in the field of congenital heart disease would disagree. The clinic's chief surgeon, Roger Mee, has the world's best success rate for the arterial switch operation. While some centers report mortality rates as high as 50 percent for this operation, most manage to keep mortality to between 5 and 10 percent, or somewhat higher for complex varieties. The very best centers have a 1 to 3 percent mortality. But Dr. Mee's centers performed 270 arterial switches for simple transposition over the past seventeen years and have lost just one patient during that time; he is noted throughout the world for his acumen generally, and for his expertise in this procedure in particular.

Bruce has already been told how lucky he is. He nods to Dr. Lane and says, "That doesn't help as much as everybody thinks it should."

Geoff, himself the father of two healthy girls, nods and looks down.

An echo machine is wheeled into the room, and Geoff does an echo of the baby's heart to assess the placement of the great arteries (they're side by side—one indicator of TGA) and measure their width. He looks at the four chambers of the heart and evaluates how well they're working, a critical factor here, since this baby's right ventricle is pumping blood to his entire body while his left is sending blood into his delicate new lungs. Geoff then checks the functioning of the four valves and examines the atrial and ventricular septums, the walls separating the chambers; often there will be a hole between the ventricles (known as a ventricular septal defect, or VSD) that allows blood to mix, but not here. He also looks for the coronary arteries, the vessels that feed the heart muscle; these are of concern because one of the biggest risk factors in fixing TGA is an unusual coronary arrangement. He gets a good view of the left artery but cannot see the right. He continues to run the probe over the baby's chest and belly, watching the screen all the while. Someone unfamiliar with echocardiography would see only a pie-shaped picture of static, but Geoff can clearly visualize the heart and at the push of a button—blue, orange, and red flash suddenly onto the screen— can measure the speed with which blood is flowing inside it, thanks to the same Doppler technology used by the local weatherman.

Bruce asks Marc what the mortality rate is for the initial procedure, the balloon atrial septostomy. Marc answers, "There's risk at every stage, but I can say that I can't remember a preoperative death from this. There's risk at surgery, and the surgeons will tell you about that. There's risk just being in the PICU." Sensing that this has not helped, he assures the father that his child "is stable and in no immediate danger."

Soon Karyn Kasnik, a twenty-eight-year-old suburban mom and occupational therapist, is wheeled through the automatic doors. It's one of the worst moments of her life. The room is full by now, with people overflowing out the door. Oh, God, she sees her pastor in the hallway—already here. She knows it: "I'm too late." She is exhausted and has begun to cry, certain that her son has died. In a minute she'll learn from Bruce that she's wrong, and she and the pastor and her mom and other family, who had a pizza party together only yesterday, will enter Connor's room. As occupational therapists, Bruce and Karyn help people recover their physical lives after debilitating events such as strokes, and so are part of the world of medicine and hospitals. That doesn't do them any good now, though; they feel helpless. Bruce, as a defense, gives arriving family members a dry account of what has happened, as if he were reciting the case history of one of his patients. The family is left alone until 12:30 P.M., when Geoff Lane, his nurse assistant, and staff cardiologist Lourdes Prieto come in to perform the septostomy, which will help Connor's blood mix.

After everyone else leaves the room, Geoff and his assistant prep the patient, set all the gear they'll need on a cloth-covered tray, and don gloves. Geoff locates the femoral vein in the baby's right groin, inserts a needle, and immediately finds the vessel. He slides a wire through the needle into the vein, then removes the needle, leaving the wire in place; he then slips a sheath over the wire and advances the sheath into the vein, then pulls out the wire. Having thus gotten a hollow sheath into the femoral vein, he can push through it a long tube with a deflated balloon on one end and a large, water-filled syringe at the other. The balloon end will snake up the vein, past the liver, and into the inferior vena cava, to enter the top right chamber of the heart, the right atrium. Lourdes, echo probe in hand, follows the progress of the tube on the screen and informs Geoff that he is in the right atrium. Geoff pushes on the tube, and it slides along the top of the atrium and then moves down the septal wall until it passes through a small hole in the septum—a hole that is open in utero but typically closes after birth. Lourdes, seeing a small line of glowing static on the screen, tells Geoff that he's crossed the atrial septum. Without these pictures, Geoff wouldn't know where in the heart he was—he might, for instance, have slipped down the tricuspid valve and into the right ventricle, or passed through the left atrium and into a pulmonary vein where he could do some real damage, or conceivably even gone in through the femoral artery instead of the vein and ended up on the wrong side of the heart altogether. Geoff squeezes the syringe to fill up the balloon at the other end. The balloon is the size of an oval marble, and nearly as hard when inflated; it fills the baby's left atrium. Geoff puts his gloved fingers over the spot where the catheter enters the groin, then flicks his wrist down sharply. The balloon pops through the atrium.

Lourdes says, "Oh, that looks good."

"It felt a little floppy to me," Geoff says of the septum, which can be muscular and difficult to tear properly.

A nurse checks the monitor and reports, "Sats look better: eighty-eight"—up from the 60s.

Geoff repeats the maneuver to ensure that the hole is permanent; it's possible that the first time, he merely stretched the existing hole. The oxygen saturation goes up to 90. Lourdes, using Doppler to judge the flow through the new hole, says, "It was very turbulent before. Now there's ample flow."

Marc asks Geoff to leave the line in the vein—he notes the size line he wants: "Use a four; if you use a five, you'll thrombose the leg"—so they'll have access through which to give the patient drugs without having to insert another line.

June Graney, one of four cardiac nurses, walks from the PICU to the lounge outside this floor's elevators to let the Kasniks know that the septostomy went perfectly—so well, in fact, that they're planning to stop the prostaglandin and let the duct close up, which should make for a speedier postoperative recovery than would be the case if they left Connor on the hormone. June is well suited to speak with grieving parents: besides being experienced, besides being a mom herself and *looking* maternal with her deep-brown eyes, her short, loosely curled brown hair, and the soft white turtleneck she wears under her blue scrubs, she can empathize, because her daughter, too, had heart surgery as an infant. She knows what it feels like when "your baby is *taken away from you*," and she can tell parents with comforting authority that this is for everyone a time of utter and terrible helplessness, as it was once for her. June will talk to the family, answer questions, ask if anyone wants to donate blood—for surgery on a newborn, Dr. Mee will use only fresh heparinized blood, less than a day old—and tell them Dr. Mee will see them sometime today to talk about the operation, which has been scheduled for tomorrow.

Joan Holden, the assistant nurse manager in the PICU, also understands what this family is going through right now: "They have to grieve the loss of their healthy child," she says.

The Kasniks spend the night at Ronald McDonald House, a special housing facility for families whose children are being treated at the clinic. After Bruce and Karyn get up in the morning—the morning of the day their newborn son is scheduled to have surgery—they drive to the Children's Hospital. The couple have not been in Connor's ICU room long when Julie Tome and two nurses appear. They must manually ventilate Connor, or bag him, as they roll his warmer bed to surgery. Karyn has not held her baby since she asked to take a nap at Lakewood a day and a half ago—her last easy minutes. The corridor is only about sixty feet long, but it seems endless to Karyn, and they stop in a foyer between the PICU and surgery so the Kasniks can formally acknowledge to a nurse that they understand what is about to happen and give their consent for the procedure. Then the doors to surgery hiss open, and the O.R. team bears Connor away from his parents. It is among the worst moments of their lives, this moment before surgery—the worst of *any* parent's life, to give up a baby to strangers who will cut open that little chest and try to fix that defective heart. Karyn wonders, "Will I see him again alive?"

About twenty minutes later, at 8:00 A.M., Roger Mee passes through this foyer and into the PICU to review films of all the clinic's cardiac patients, along with a dozen other docs and nurses. Afterward the group will round on all patients both in the PICU and on the floor, where recovering, no-longer-critically-ill patients are sent. Mee's stature is modest—he stands just under five feet six—and his dark tweed jacket is buttoned over a solid gut that adores an egg-bratwurst-and-potato breakfast. The band of hair surrounding his bald, flat pate is dark and trimmed close, and every day he wears the same thing to work: jacket, black slacks and laced black shoes, white shirt, and one of two ties.

During rounds Mee, a fifty-six-year-old native of New Zealand, is the silent focus; young resident docs wait for his attention before beginning each patient's presentation. Mee will get back to his office between nine and ten, depending on the number of patients in the unit. Between operations yesterday—an ascending aorta repair, an aortic root replacement, and a pulmonary artery band—he spoke with the Kasniks about the operation he would perform on their child, telling them that it carried a 2 percent risk. He said to them what he says to all parents almost every day, emphatically: "Risk *means:* death or near-death resulting in severe brain damage." He explained that the main risk factor was unusual coronaries and said he wouldn't know what Connor's were like till he saw his heart. He told them he'd lost only one out of about 270 straightforward switches (a 1.8-kilogram baby, more than a decade ago). He added that about 3 percent of these patients required another operation within their first three years to repair a leaky valve, a residual ventricular septal defect (VSD), or pulmonary or aortic stenosis. The long-term results, he said, seemed to be good. Long-term problems he'd seen included irregular heartbeats, ventricular failure, and valve problems. Most kids who underwent this procedure spent an average of three days in the PICU and another week on the floor before going home. Given a choice between being grim and being hopeful in tone, Mee always chooses grim.

The Kasniks found Dr. Mee neutral and thought he was probably distracted—"not warm and fuzzy," Karyn will later recall. They registered nothing bad about him but nothing comforting either. The one moment Bruce clings to, the moment his heart lifted, came when Mee said the words "two percent risk"—a statistic far better than what Bruce had imagined. Two percent—he could almost allow himself to hope for the best. After the meeting Mee wrote down and dated exactly what he'd spelled out for the parents, including the risk he'd quoted and the possible surgical complications, such as bleeding and nerve damage.

On this Friday morning, as Connor's chest is being opened up, Mee hasn't been back in his office long when a secretary passes on his fifteen-minute warning. Normally Mee might have stood outside his office after rounds and talked with Debbie Gilchrist, his secretary and the coordinator of the department, about work-related matters—plane tickets to conferences, phone messages, consultations—but Debbie is off this week, and he didn't stop today to speak with the temporary secretary. Typically he'll sit down at his cluttered desk, finish his dictation, take a call from a referring doc, and meet with the line of nurses and cardiologists who are standing outside waiting to speak with him about various patients. But this day, unusually, no calls come in, and the inevitable queue does not form. Alone and unbusied by his standard morning tasks, he decides to dress for the O.R. He opens a tall white cabinet across the room from his desk and changes into his operating uniform: white trousers, white short-sleeved shirt, white bucks. In a few minutes he'll reach into his jacket pocket and retrieve another square of Nicorette, which he'll grind away at all morning long.

The secretary calls to him, "Dr. Mee? They need you in the O.R."

Often the O.R. will call again after the fifteen-minute warning just to hurry him along. Mee gets sidetracked by so many people that it can take him ten minutes or more to get there—even longer if he stops in to see parents who want to speak with him one last time before their child's operation. Had Debbie Gilchrist been here today and taken this particular call from Lori, the O.R. circulating nurse, she might have said the exact same words as the temporary secretary, but she wouldn't have called them out. She would have risen from her desk and strode around the partition, saying "Dr. Mee" as she hustled, then stood in the door till she caught his eye and said quickly, *"They need you in the O.R."* He would have known to break off all conversation, leave his office, hurry down the hall, and enter the surgery adjacent to the PICU, a twenty-second trip. But this secretary, temporary and unaware of the urgency that has been relayed to her by the circulating nurse, merely calls out, "Dr. Mee? They need you in the O.R." Nothing more.

Because he is already dressed, because he is not on the phone or speaking with a colleague, he leaves his office for the O.R. without delay. He walks down the corridor as he usually does, hits a square plastic button high on the wall that opens the doors to the two operating rooms and the surgical lounge, and retrieves a hood and mask. He puts these on, ties them, tapes the mask across his nose, then from a cupboard casually removes two wooden boxes, one containing his louped glasses, the other a fiber-optic headset. He puts both on, then walks into the O.R., plugs in his light cable, holds his hand before him where his surgical field will be, focuses the beam, unplugs the light, and leaves the room again to scrub.

From his position at the operating table, Mike Fackelmann has a view of the O.R. door, and when it slides open with a hiss, he looks up to see Dr. Mee. At first he's relieved, but then he doesn't understand—*What's he doing, dicking around with his headlight?*—and when Roger, unbelievably, *leaves,* Fackelmann puts down his instruments—*It takes six minutes to scrub,* he thinks; *by then the kid will be dead*—and strides to the door, which has not yet hissed shut. *"Roger, we got a problem,"* Fackelmann says. *"We need you in here now."*

Roger doesn't even scrub. Hearing that they're in the duct, he's gowned and gloved in moments and at the table with a finger on the hole and saying to Bob the scrub nurse, "Stitch, please," and then he tries to suture it closed beneath his finger, even as blood fills up the chest cavity and obscures his vision, Mike trying to suck it all out so Mee can see.

Mee says, "Jesus, I've never seen this before." He quickly puts in three 6-0 Prolene sutures diagonally across the edge of the hole, but it's not working, it's like sewing tofu, and so he has to keep his finger on it while they get this kid on pump fast. It's anything *but* fast, though—not only are they in an urgent situation, they're without one critical hand, Dr. Mee's, and it throws the whole routine off. Visibility is a problem because of the fairly massive bleeding, the chest cavity bubbling up with thick red blood. The oxygen saturations are only 40 percent; it'll be a race to get this child on pump before his heart arrests. Frank Moga, a third surgeon, like Mac a thirty-five-year-old fellow, steps up to the table beside Fackelmann and across from Mac and Dr. Mee, having scrubbed as soon as he learned there was a problem, and wanting to help if he can.

With his free right hand, Mee makes a small hole in the ascending aorta and inserts the cannula, a tube with an L-shaped steel tip. This will shoot oxygenated blood from the heart-lung machine into the aorta and to the brain, to the arms and down to the lower extremities of the body, to all the organs. It has a little white ring around the end that helps fix the tip inside the vessel. It's almost impossible to get it in properly with one hand: as soon as Mee releases the cannula, it pops out again. "Oh, shit!" he says. He pops it back in and says to Moga, "Please, snug it up quick." He says, "I've never *seen* this before."

Then Julie's voice: "He's arrested."

"Give the steroids," says Mee calmly. Mee believes that steroids, a powerful antiinflammatory agent, will help protect the brain during a traumatic event. He repeats, "I've never seen this before." And they continue to work on getting the venous cannula into the right atrial appendage, where it will suck in all the blood returning to the heart and send it to the oxygenator. The heart resumes its beat on its own, and Julie quickly says, "Sats are forty, we've got a rhythm back; venous is thirty, our pressures are back."

With the aorta cannulated at last, Mee says, "I want another blood gas."

Julie says, "I'll get you another. Our blood gas during the arrest was good."

Addressing George, the perfusionist, Mee asks, "What flow are we?"

George says, "What *flow?*"

"Oh, *God,*" Mee says, disgusted. "Go on bypass, come on!"

And then, finally, the patient is on. Mee mutters, "If this were a seven-forty-seven, we'd never have left the airport." He asks, "What's the pressure, please?"

"Thirty," says Julie.

"Come up on your flow. . . . Start to cool."

"Cooling."

"Can you come up on flow?" Mee says, still not liking the pressures. George can raise the blood pressure by increasing the amount of flow, but only to a point—too much resistance can indicate a problem, or even create one.

"I'm afraid I'll overload him," George responds. "I don't have the drainage." He ups the flow a little, then says, "The aortic resistance is high."

Mee says, "I don't know if the damn thing's in." He fusses with the aortic cannula. "That's what I'm worried about. We're not explaining the high resistance." He fusses some more and then asks, "Better, George?"

George says, "Two-ten. I can live with that."

Mee begins to work on tying off the duct. A fountain of blood shoots straight up out of the patient's chest. Mee says, "Christ," but never interrupts his deliberate movements. "Is there a hole in the aorta?" he asks, trying to figure out why the chest is filling with blood. He ties off the duct at both ends with 2-0 silk. "Give us another stitch, please," he says. Bob has one ready and slaps the needle holder into his hand. Mee checks the aortic cannula to make sure it's secure, looks at the monitor, puts his instruments down.

"I've got to go get scrubbed," he says. He steps away from the table and exits the O.R. to prepare for the most critical part of the arterial switch operation.

The table is silent for a moment, until Fackelmann says, "Take a break, Mac."

Mac is silent.

Frank Moga concurs with Fackelmann, advising in a friendly, consoling tone, "Go talk to him."

Makoto Ando nods and leaves the table. Mac is fully responsible for this incident and he feels every ounce of the blame. He goes out to accept his fate from Mee.

Mee is parental, even breezy as they talk, standing at the scrub sink immediately outside the O.R. If Mac hadn't come to him, had tried to ignore the problem, Mee might have torn him to pieces— but as long as his staff are honest with themselves and with their colleagues, Mee sees no need for abuse. He knows Mac to be a good and conscientious surgeon. He seems casual and unconcerned. "One thing you never do," he tells Mac, all the while scrubbing his nails, fingers, and arms with a sudsy brush, "is pick up the duct." He shakes his head, as if he still can't quite believe it.

When they're both back at the operating table, Mee inserts into the root of the aorta the small tube through which they will send the cardioplegia—a potassium solution, poison—that will pass through the coronaries feeding the heart muscle and stop the heart's beating. The heart will then become flaccid and pale.

Mee says, "Plege on." He rests his hands atop his gut and stares at the heart, waiting for the plege to take effect. He shakes his head. "Good thing the coronaries are the sensible ones." He pauses but continues shaking his head. "What you might have seen is some sort of bulge there. The neonatal duct, nobody should touch. Never touch anything that you don't know what it is. I've never seen that before."

In the third-floor parents' lounge, the Kasniks wait. Furnished with a long table, couches and chairs, and a television, with another, private room for calls, the lounge is comfortable but devoid of personality. The couple are sitting directly under the O.R. in which their son is being operated on. When June Graney comes through the doorway for the first time, they turn to her. She will be down throughout the operation to keep them informed of Mee's progress. She tells them that there were some problems with bleeding, but they've gotten it under control. A "glitch," she calls it.

When the cardioplegia has been running for four minutes, Mee asks for scissors, and the switch operation begins. Mee prefers not to operate on neonatal hearts, largely because at this stage the heart muscle, or myocardium, has yet to mature into its more solid state, and water-rich neonatal myocardium is harder to sew. It's a tiny thing, the heart of a six-pound baby, and that, too, increases the risk. Also, Mee believes that a newborn is more vulnerable to the insults of the heart-lung machine and anesthetics. But there's no choice in this case, because the septum between Connor Kasnik's ventricles is intact; having pumped only to the lungs until now, the left ventricle won't be strong enough to pump to the entire body if they wait.

Surgeons began trying to repair this particular heart defect in the 1950s, but with only moderate success. Until the mid-1970s, various techniques were used that had a low early mortality rate but a high incidence of subsequent sickness and death in the teenage years. In 1975 the first successful arterial switch was performed. While the long-term results would prove promising, the surgical mortality for this new operation, the switch, was increased, presenting parents of TGA babies with a brutal choice: live infant, dead teenager, or high risk of death now in exchange, maybe, for a longer life.

The idea of the operation is simple: there are two tubes rising out of the heart next to each other, and they get switched. The problem is those little coronaries, the key to success or failure. They usually come out of the left and right sides of the aorta and branch out over the left and right sides of the heart, feeding the hungry muscle. But they don't like to be moved. They tend to kink, and when they kink, no blood goes through them, causing what is in essence a heart attack.

As the 1970s progressed, surgeons learned how to safely transfer two coronaries from the aorta to the pulmonary artery. In January 1983 surgeons at Children's Hospital in Boston became the first to perform successful switches on neonates. Roger Mee, then working at the Royal Children's Hospital in Melbourne, Australia, did his first switch in April 1983. He would prove to have a special finesse with and understanding of the coronary transfer and would publish widely on strategies for handling even the most difficult mutations (single coronaries and intramural ones, which bend around inside the wall of the aorta). So when Mee calls the Kasnik baby's coronaries "sensible," it's no idle compliment: bad coronaries are the chief cause of mortality in this surgery.

Before snipping the aorta just above the coronaries, Mee uses small sutures to mark where he wants the coronaries to be attached on the pulmonary artery, providing a general idea of their height and placement. When the heart stops and all the blood drains from it, it goes flaccid, flattens out, and a surgeon can lose his place. Mee bisects the aorta and snips the coronaries out, leaving a big cuff around the tiny opening so each one looks like a little button, about a quarter inch square, affixed to a little artery whose opening in the center of the button is about the size of a pinhead. Mee next bisects the pulmonary artery, but a couple of millimeters higher up than the aorta, and makes some minuscule L-shaped cuts near the pulmonary valve; he then sews the buttons into these new slots with sutures that are so fine as to be barely visible, 7-0 Prolene in a running stitch. It's like sewing in a dot of tissue paper—hence his need of loupes. He then sews the main pulmonary artery to the aorta.

Next Mee has to repair the hole Geoff Lane made yesterday with the balloon septostomy. He takes the patient entirely off bypass (to prevent air from rushing into the venous cannula and shutting down the bypass machine) and cuts open the right atrium, where he finds a big open tear in the inner wall, which he sews up with a running stitch. He then closes the atrium and puts Connor back on pump.

In the four minutes it has taken Mee to do all this, the circulation to the patient's brain and body, cooled to a temperature of 22 degrees Celsius, has been stopped, placing him in a state called circulatory arrest. Mee now completes the operation by patching the holes left by the transferred coronaries and sewing this aortic root to the pulmonary artery.

The procedure concludes with the insertion of temporary pacemaker wires, lines in the heart itself to measure pressures in the various chambers, a double lumen catheter, a Tenckhoff catheter in the abdominal cavity, and chest drainage tubes. After Connor is given dopamine (a drug to help his heartbeat) and nitroglycerine (to ease vascular pressures), the cannulae are removed.

The baby's heart begins beating on its own, and soon he is stable, with good pressures. He has been on the heart-lung machine for two hours and twelve minutes, with his heart stopped for forty-two minutes of that time. The right ventricle now pumps blood through an aortic valve into the lungs. The lungs return the blood to the left side of the heart, and the left ventricle pumps it out again through a pulmonary valve, into the aorta and then out to the body. The switch has gone perfectly, and Connor has responded accordingly.

Roger Mee says "Thanks very much" to his assistants and steps down off a low footstool, away from the table. Mac and Fackelmann will close the chest. Lori unplugs Mee's headlight, and he takes off his gloves and gown, signs his name on a sheet of paper attesting to the procedure he has just performed, and leaves the O.R. Once outside, he removes his headlight, peels off the mask and paper hood, and says, "Jesus *fucking* Christ!"

His face is red and creased bright pink from all his headgear. His eyes are wide and his expression is crazed, even giddy. He almost looks like he's about to break out into laughter. His partner, Jonathan Drummond-Webb, a tall South African who began here as a fellow five and a half years ago, at age thirty-six, happens to be standing outside the surgical lounge, leaning on the front desk. Mee tells him what happened and asks, "Have you ever tried to cannulate with one hand? I had my finger on the duct, and I had to put the cannula in with one hand, and each time they took it, it came out!" He returns his headlight and glasses to their wooden boxes and places them in the cabinet. "It's all in doing the details exactly the same way. I keep trying to tell these guys that. It's not by accident I do it; I figured it out a long time ago." He pauses, shakes his head, then starts in again on "these guys," the fellows: "They get paralyzed when something like that happens," he says. He's got two more kids to do today, an arch repair and complete repair of tetralogy of Fallot. He chats a bit longer with Jonathan, then heads back to his office to have some lunch, still shaking his head. Jesus fucking Christ.

2. The Virtuoso

I'd been hanging out at the Center for Pediatric and Congenital Heart Diseases at the Cleveland Clinic for two months when the shell-shocked Kasniks arrived. It had been a busy time for the center's staff, filled with all kinds of drama and tragedy, deaths and saves, errors and grace, and mostly a lot of work and stressful nights on call—the routine stuff of intensive-care units, critical-care medicine, and peds heart centers across the country. On one of my first days, while I was speaking with Roger Mee in his office, a call came in, and he answered it and then said to me, "Arrest in the cath lab," with little more urgency than he might exhibit if he were commenting on the weather. Just before jogging off, he paused, turned back to me, and asked, politely, "Do you want to come along?" His partner, Jonathan Drummond-Webb, had left his O.R. patient to Frank Moga and raced down there before Mee, the cord to his headlight trailing behind him like a banner; by the time I got there, he had already begun mighty chest compressions in an effort to resuscitate the patient—a teenager who'd been undergoing a catheterization procedure, and whose feet now lifted off the table with every push—successfully, as it would turn out. The last time this had happened—happily, it was a rare occurrence—the patient was a baby, and Mee had had to open its chest right there on the cath lab table.

There had been a lot of sick babies at the center during those two months, and the trend would continue all the way through De-cember, with all manner of the worst kinds of heart problems a baby could survive; in three instances the problems were so bad they had to put in new hearts altogether. But this one case, that of Connor Kasnik and family, was a microcosm of the world of peds heart sur-gery, an emblem of the progress that's been made in the field of con-genital heart disease—and not just in open-heart surgery, which began in the 1950s, but more recently, through the 1980s and 1990s, in echocardiography, interventional cardiology, cardiopulmonary bypass, and neonatal surgery. The repair of this particular baby's heart defect was an elegant description of congential heart disease generally, not least because it was a success, as is most neonatal heart surgery today. Connor was extubated the day after the switch, was taken off the ventilator, and soon awoke; when I saw Bruce, he looked rested and had a big grin on his face. "It's a *great* day," he told me, brimming with exhilaration. Exactly one week later, Connor lay in a crib in the ward, wearing jammies and covered by a red blanket shaped like a Christmas stocking—"what he was supposed to go home from the hospital in a week ago," said Karyn, evoking the powerful joy-sorrow that most parents here feel. ("I would never call this a cure for life," Mee says of the neonatal switch operation.) Above her son, a blue Elmo propped below a crucifix decorated the curtained-off area, along with "It's a Boy" and "Get Well Soon" bal-loons and four wooden snowmen with "Mommy," "Daddy," "Kyle," and "Connor" written on them. The Kasniks were preparing to take their baby home. This was the norm at most big heart centers.

Finally, Connor was an emblem, for me, of a remarkable truth that never left my mind in all the time I spent in the O.R.: when a baby or a child or an adult is on the table with his or her chest open, disaster is never more than a breath away, no matter how rou-tine or simple the case may be. One small breath.

None of this, however—not the medical advances, the drama of the O.R., or the pathos of the families' stories—was why I'd asked to be here in the first place. I'd originally come to learn more about the craft of surgery and better understand the people who make it their daily work.

Story ideas evolve from all kinds of sources, and this one had its beginnings in a conversation over dinner with a family friend named Stu Eilers. Stu is a big guy with a big voice and a dramatic sense of story that serve him well in his courtroom appearances as a civil defense lawyer. He had a friend I ought to meet, he said—a heart surgeon. A *pediatric* heart surgeon, to be precise. Roger Mee operated on babies' hearts every day, Stu explained, and was reputed to be one of the best such surgeons in the world. Stu, as usual, had some choice stories to tell in support of his claim that this Mee fellow was indeed worthy of interest. There was the time, for example, when the Israeli government "kidnapped" Mee while he was on vacation in Italy with his wife, flew him to Israel, and all but forced him to operate on a dignitary's child; the time he was flown to Saudi Arabia to perform surgery on the king's grandson; and the time when, after word got out that he was considering moving to the United States, two men from the Cleveland Clinic (operatives in dark glasses and black trench coats, in Stu's portrayal of the scene) arrived in Melbourne, Australia, to make him a generous offer.

All this made for excellent dinner chat, but it wasn't what hooked me. The thing that stuck with me after that evening those few years ago, when I first heard Roger Mee's name, was Stu's simple comment, uttered with almost hilarious disbelief: "I mean, what a *job. Imagine* it."

What a job, indeed. This *was* interesting to me. Here was a man who operated on babies' hearts for a living most days of his life, often doing two or three operations in a single day. And he was not just *good* at fixing hearts that were sometimes no bigger than a walnut; he was one of the best in the world. For a quarter century he'd been doing this. Imagine the stress of that work, Stu had said. Look at the questions bound up in it: What child *don't* you try to save? What if a patient can't pay, what do you do then? Not operate? And if you do operate, how many free operations do you do? How do you get to be the best in the world at one of the most difficult surgical specialties there is? Who do you become when you do this work? And how do you respond to the job's ultimate responsibility: the fact that your personal skill and intelligence, or lack thereof, will determine the quality of this child's life from there on out, may decide whether he or she lives or dies, will change the lives of the parents who have entrusted you, a stranger, with their most precious treasure? Who *was* this guy Mee, anyway?

At the time, in the winter of 1999, I was about to move my own family to the East Coast, where I would work at a boatyard in Vineyard Haven, Massachusetts, and learn more about the people who build boats out of wood, craftsmen and sailors whose livelihood is based on their ability to bend planks around frames. For the three previous years I'd been writing about professional cooks and the work of cooking, had even for a short time became a clock-punching cook myself. I wrote about people at work; work was important. But not just *any* work or anyone at work: I tried to concentrate on those few who were considered by their peers to be among the very best in their professions. I'd developed at cooking school a fascination with people who were exceptionally skilled at what they did, had discovered that people who pursue perfection, date-on-a-dime clarity, and impossibly high standards are the most compelling human beings alive, and I'd decided that if I was going to spend my life watching and writing about people at work, I was not going to focus on what was merely average or, worse, mediocre. I wanted to see the best. By chance or by unconscious design, my chosen subjects were people who used their hands for a living, people who worked with natural products, food and wood. And now my friend Stu had planted in my mind a vision of a different kind of laborer, a man who like the others worked with his hands for a living, another genuine craftsman, but one whose "product" was living tissue and whose trade was plied on the hearts of children, often while those hearts were beating.

I'd interviewed the writer David McCullough for my wooden boats book, and he'd articulated a truth I'd already seen with chefs and was witnessing again with boatbuilders: "We're shaped by how we go about earning our daily bread," he'd said. "We *become* what we do. We are shaped by our choice of vocation and what demands that puts on us, what expectations, standards—ethical, professional."

This was surely true of the builders of wooden boats: they *were* their boats. A certain integrity, economy, and resourcefulness defined both their lives and their work. I'd spent nearly a year at the most prominent cooking school in the country, where more than a hundred chefs regularly teach students what it *means* to be a chef. Amid this concentration of talent, I'd seen unmistakable patterns ranging from near-militaristic perfectionism to apparent madness, qualities that seemed to arise out of the intense demands of the work; chefs, I realized, were lunatic with enough frequency to suggest that the work *made* them that way. If this was true, who, then, did someone *become* who operated on babies' hearts every day of his life?

Late in the winter of 2000 I called Stu and asked if he'd introduce me to Roger Mee. He agreed and dropped by Mee's house one weekend morning to explain who I was and why I wanted to speak with him. Roger didn't much like the idea of being written about. He recalled for his wife, Helen, the insufferable press coverage that had followed his first heart transplant—the first such operation in Victoria, Australia, in fact, that was a long-term success—and the opening of that country's first and only pediatric transplantation center, whose services included heart-lung and neonatal transplantation. Both Roger and Helen came from humble backgrounds— Roger was born near what's now Quetta, Pakistan, to an Irish army-garrison chaplain and a teacher from New Zealand, while Helen was raised on a New Zealand sheep farm—and both were deeply modest, self-effacing people. I've tried to appreciate what it's like to be written about, and I can't imagine it's completely comfortable (I'd have sided with Mee had I been his friend). Helen and their daughter, the youngest of the four Mee children and a senior in high school, Stu recalled, were reportedly intrigued by the novelty of what Stu had proposed, but all remained apprehensive about allowing a writer into their lives.

The Mee residence in Gates Mills, one of Cleveland's most exclusive enclaves, was so big, Helen had left a note on the front door telling me to let myself in; from the kitchen, they wouldn't hear me knocking.

Roger, dressed that day in a blue chambray shirt and khakis, was and is a handsome man, with light-blue eyes and an easy laugh. He was a boxer in school and still commands that kind of attention through his straightforward gaze and direct comments. We sat down to talk. He explained his overarching goal simply: "to provide the safest possible care for children and adults with congenital heart disease—'safest' meaning minimizing the mortality—and provide that for as many people as we can possibly manage."

Congenital heart surgery, which not too long ago was a last-ditch, win-or-lose long shot to save a patient's life, is now all about safety and reducing risk. "Our biggest contribution, really, was the wherewithal to do much safer open-heart surgery on babies," Mee told me, referring to his work in Australia. "We seemed to make a big breakthrough there in terms of risk—because of a bunch of things in preoperative care and postoperative care, but in particular a different view of what sort of blood flow babies needed and how to get it into them without blowing their vessels apart."

For most of its early history, heart surgery was the story of maverick doctors and what today would be deemed appalling mortality rates. "The first surgeons were swashbucklers with a virtual license to kill," Mee said. "And that's how they learned. Everything I do, I can do because of what's come before me."

Together, Roger and Helen recounted a little of their own history. Both educated in New Zealand, they'd met in college and married when Roger was in medical school. By the time he was thirty-three Roger had already worked under two of the world's most renowned heart surgeons, Brian Barratt-Boyes in New Zealand and Aldo Castañeda in Boston. He made a name for himself in the 1980s at the Royal Children's Hospital in Melbourne, where his center recorded mortality rates so low that many people thought he must be lying. In the early 1990s the University of California in San Francisco contacted him to ask if he would consider taking over its peds heart center. He turned down the invitation, but it got him and Helen thinking: Did they want to live out their years in Melbourne, maintaining what he'd worked so hard to create, or should they build a new life somewhere else? They almost always opted for building, and so entertained offers from some thirty centers in the United States. In the end, Texas Children's in Houston looked like a good prospect, as did the Cleveland Clinic. A surgeon named Toby Cosgrove had developed one of the world's most successful adult heart programs in the latter, but this famous medical behemoth had only a small congenital heart program that could claim no better than average results. The Mees chose Cleveland for a variety of reasons, not least of which were the promise that Roger would have carte blanche to build a proper pediatric heart center and, doubtless, a generous salary. (When I asked what his salary was, he told me it was "private," in a tone meant to end further questioning, then grew annoyed when I pressed. He guessed that chief peds heart surgeons could earn anywhere from $400,000 to as high as $2 million in some places. A seven-figure salary for a surgeon of his reknown would not surprise me nor seem remarkable.) And so the family moved to Ohio in 1993.

There was considerable resistance to his demands once he actually got to the clinic, Mee recalled. For one thing, the carte was not quite as blanche as he'd been led to believe. Then, too, there was an extraordinary amount of politicking going on in the service of power and ego—a dynamic that is never in the interest of patients, he noted. When it became clear that Mee was on the verge of departing in anger, however, the CEO, Floyd Loop, spoke with him, placated him, and helped him negotiate the bureaucratic tangles.

Loop, as the chief executive officer of a medical institution that generates more than $3 billion in revenue annually and is ranked third in the country by *U.S. News & World Report,* is a powerful guy. He was for decades a heart surgeon himself, at what is now judged to be the nation's top heart center. In courting Mee, the clinic's board had obviously been seeking to augment its staff with someone who had an international reputation, a star. But Loop says that with this particular hire, the board got even more than that.

"He exceeded my expectation in the first week, and then the first month and then the first three months he was here," Loop reminisces. "The first week he was here, people from all around the world, particularly the Far East, came here for the big operations, switches and things like that. It's very hard to command that kind of following where people will come from everywhere. And then I looked up and a few months had gone by, and this guy had rattled off a hundred, a hundred fifty cases, with no mortality. And I thought, Ya know, we really got something here. I remember one year he did something like five hundred cases, with two deaths or something. You don't find those statistics everywhere. These are big cases.

"I've always suspected that the really great congenital heart surgeons," he continues, "are really great because not only do they have the gift of technical ability, but they've gotten better as they've gone along and gained experience. And then more and more patients come to them, so they gain more experience. And so there are only a handful of those guys who are really world-class surgeons who have seen and done everything. In adult surgery there are dozens and dozens of really great surgeons. In pediatrics, because of the small pool of patients, only a few people really acquire the experience."

Mee himself, bored by the typical stories of surgeon-as-savior, preferred to talk to me that first day about the craft of surgery rather than his results. He liked to build things, he said proudly, such as the family's beach house, an A-frame on a remote stretch of sand in New Zealand; surgery required that same kind of physical skill. A person had to be naturally handy to be a surgeon. (At age five, Mee once told an interviewer, he fixed his parents' washing machine by jury-rigging a new axle. When asked about that story now, he says he must have been seven or eight, and he underscores the point that he *needed* to fix the axle, which he did with some wood and ball bearings, because he was the one who'd broken it in the first place—"An old Bendix washing machine," he recalls wistfully. "Ran for ages.") And being a good peds heart surgeon demanded a high level of craftsmanship indeed.

"It's a lot like any artisan craft," he said. "It's not a robotic skill—if it were, you could train anybody to do it. In surgery, you have to accommodate all the variations of nature, and in the case of abnormal hearts, you have an infinite variety."

The body was more likely to tolerate good craftsmanship than bad, he said, and the difference was especially salient when it came to a child's heart. "If you do things accurately, well," he explained, "without bruising, stretching, or damaging tissues, it's much easier for the body to recover. You need precision, and you've got to have good judgment on just how much of this incredibly valuable, rapidly growing tissue you can strangulate or destroy or cut out. How far apart the sutures are and how you preserve blood flow—it all contributes to the healing process."

Before that day, I'd known next to nothing about congenital heart disease, and I'll bet the same is true of all those who either aren't in the field or haven't, like the Kasniks, found themselves horribly in its midst. Even the name had a distancing effect: heart disease. But is it really a disease? It's more accurate and clearer to say "defect." Usually when people talk about heart disease, they're referring to coronary artery disease, which is an "acquired" condition; "congenital" heart disease is a condition existing at birth.

Congenital heart defects are the most common form of birth defect overall, affecting nearly 1 percent of all live births, and the ones most likely to result in death, according to the American Heart Association, which works closely with the National Center for Health Statistics to compile its yearly statistical update. Of the approximately 4 million babies born each year in the United States, some 32,000 will have a heart defect, and in most cases there won't be any way of knowing what caused it. Of those 32,000, between 19,000 and 26,000 will need some sort of intervention or surgery. Thousands more are born each year with defects that go undetected. Thanks to enhanced diagnostic tools, people who were born before the 1980s are increasingly having surgery to fix "tolerable" defects—for example, arrhythmias—they have unknowingly had since birth; indeed, there will soon be more adults operated on for congenital defects than infants. About 1 million people live with a congenital heart defect of one kind or another in the United States. Exact figures are not available on how many total operations are performed each year to repair congenital heart defects, but the count for open-heart surgeries just on children under fifteen is 25,000 annually. The estimated cost of all in-patient surgery exceeds $2.2 billion. In 1999 heart defects and their surgical treatment claimed 4,436 lives, the majority neonates, infants, and children— but even that number reflects a sharp decline in mortality rates, largely as a result of better diagnostic techniques and improved surgical and intensive care. In the 1970s roughly 30 percent of those who underwent surgery to repair heart defects died; today the figure is around 5 percent, though success rates vary widely among centers that offer the surgery.

The problem is even more serious in third-world countries, where hospitals are typically poorly equipped to deal with congenital defects. Compounding the difficulty is the fact that, as Mee notes, the birth rate in such countries is three times as high as the U.S. rate, and a much smaller fraction of those born with heart defects can be treated.

The number of cases of congenital heart disease is not insignificant, but the total is dwarfed by comparison with acquired heart disease, about 80 percent of which is coronary artery disease. Fourteen million people suffer from clogged coronaries, which are to blame for about one of every five deaths in the United States, or some 500,000 deaths each year, more than are caused by any other single illness. The management of these arteries consumes about $90 billion in health-care funds annually—again more than the cost for any other disease. The adult cardiac unit at the Cleveland Clinic performs about 3,700 surgeries per year, some 34 percent of which are coronary artery bypass grafts, a procedure abbreviated on preop cards as CABG and routinely referred to as "cabbage." (Unusually, half of the clinic's cardiac procedures in 2000 were valve replacements and repairs, as opposed to cabbages.) Both the numbers and the dollars involved in bypass grafts are huge, which may explain why this is where most people's knowledge of heart "disease" begins and ends. Most people have never heard of transposition of the great arteries, tetralogy of Fallot, coarctation of the aorta, or the insidious hypoplastic left heart syndrome, to name only a few of the thirty-five documented variations of congenital heart defects. The phrase "heart defect" tends to conjure up nothing beyond a vague notion of a hole somewhere, as if the heart were little more complicated than a pair of pants. But in fact the variations of defects—which can have to do with the shape of eight main vessels, where they are connected, valve malformations, problems with the muscle itself, the shape of the chambers, or the conduction system within the muscle—are infinite, since even similar defects will vary from heart to heart and since many variations are a mix and match of several different lesions. Many hearts, moreover, are so peculiar and individualistic as to defy any attempt at categorization; these are simply referred to as "complex," or alternatively, in the words of one cardiac nurse I met, "completely fucked up."

The more I read about the defects to which the heart is vulnerable, the more apt that nurse's description sounded. The defects seemed on the surface so bizarre that I had to wonder, how could such hearts possibly survive gestation? Defective hearts get along beautifully in the womb, it turns out, happy as demented little pearls.

The normal heart—its structure, its double pump action (the atria pumping blood into the ventricles, and the ventricles pumping it out), its electrophysiology—seemed a small miracle. The cells of the heart are unlike any others in the body; they form an interconnected net, unlike striated or smooth muscle, so a beat anywhere in the heart is communicated throughout the organ. When it beats properly, the contraction is not a uniform squeeze, not a simple contract-relax mechanism, but rather a squeeze that moves up the heart like a line of dominoes toppling over. The atria are thin, almost billowy sacs resting on pure, powerful muscles, the ventricles, which beat perpetually. It's a beautiful but fundamentally primitive scheme.

From the moment its first cells twinkle into being, cells generated by the embryo's hunger, the heart is a feat of evolutionary genius. Just a couple of weeks past fertilization, these cells will have already formed a tube that pushes blood through the cell cluster. At about a month, this tube performs a critical maneuver: it pretzels, it loops and twists. "That's when it really gets interesting for me," says surgical fellow Frank Moga. "That's when it starts to show its real complexity, when it makes that turn. It really achieves its three-dimensional beauty." It's also where things can begin to go wrong.

In addition to looping, the nascent heart must also now start to separate within itself into a right side and a left side, and each of those sides must have its distinct incoming and outgoing channels. Sometimes, these outgoing channels—an aorta-pulmonary helix—don't twist all the way, so that the aorta and pulmonary artery emerge from the wrong ventricles (transposition). And sometimes, as the single muscular ventricle and the single thin-walled atrium squeeze off to form two of each, the inner wall doesn't quite grow together, leaving a hole between them (septal defect).

At the same time, two wispy valves—drifts of feathery-looking tissue, fine as a butterfly's wing—should be forming between the atria and ventricles, as should two other, stiffer, three-flapped valves within the pulmonary artery and aorta. Any of these complex structures within the fetal heart can get stuck or not form at all. And all the while this organ must continue to squeeze nutrients through itself and therefore through the fetus. After four weeks, it's gotten big enough to demand its own fuel supply, and in response a network of arteries has developed all around to feed it: the coronary arteries, which sometimes grow in odd places and occasionally come out of the wrong vessel altogether. As the arteries are developing, the other critical network is likewise taking shape: complex series of specialized muscle cells—cells that transmit electrical impulses—gather in the right atrium, then channel through muscle to another node in the ventricular part of the heart and disperse from there. These cells signal an atrial contraction, forcing blood down into the ventricles; then an electrical charge signals the ventricle to initiate its own contraction, propelling the blood up and out.

That's the heart; it's complete. The little creature growing around it is not even a fetus yet (at eight weeks old, it's still called an embryo), and this ingenious pump—four chambers, four valves, arteries, veins, and a conduction system—is the size of the head of a pin.

Complex and entire though it is, however, because the little fetus receives oxygen from its mother's placenta and not from its own lungs, the heart remains a single-system pump, replete with a variety of holes and passageways (such as the ductus) that will close after the baby is born. As long as it keeps squeezing, the fetal heart can have any number of strange deformities and still function perfectly well—that is, until the sea change of birth, when vascular pressures flip-flop, the lungs fill up with air and those holes and passageways shut down. The electrical-biological-mechanical contraption that is the human heart is so complex in its design that unhelpful variations in its intended form—heart defects—occur more frequently than do anomalies in any other part of the growing fetus. And the trickiest part is that this most common of all birth defects often remains utterly hidden until a baby is born.

It's at that moment, when the heart begins its lifelong double business—pumping blue blood to the lungs and red blood to the body—that most congenital defects start to wreak their havoc, sometimes with mortifying speed, sometimes wickedly slowly. A hole between the ventricles, for example (known as a ventricular septal defect, or VSD), is a common problem that can usually, if diagnosed early, be quite easily repaired; undiagnosed, however, it can over months or years cause terrible lung and heart damage that may be irreparable. The worst of the defects, such as those that result in single-ventricle hearts, require extraordinary ingenuity to fix—indeed, surgeons have only recently come up with ways to jury-rig these defective pumps and keep babies who have them alive.

For most of human history, about half of those with congenital heart disease died of it. Then, in the middle of the twentieth century, maverick surgeons with a "virtual license to kill" (in Mee's description) figured out how to fix some of these defects, whether by closing holes here or creating new vessels there, and as time went on and technology and knowledge advanced, they discovered they could put in valves and revise the whole structure and ultimately, in the 1980s, even replace a bad heart with a good one. The congenital heart was now a puzzle that could be understood, fixed, and managed through skill and ingenuity, artistry and intuition. The tools used to diagnose this stuff—once it was a stethoscope, then came rudimentary echo, which afforded a one-dimensional view of the heart—have transformed the surgeons' work and their patients' lives. Today's cardiologists have two-dimensional echo, which provides clear views of all chambers and vessels, and Doppler ultrasound, which lets them visualize and measure the velocity of blood moving inside the heart. For exact pressures and precise pictures they have angiograms, movies of the blood flow through the heart. They thread a tube up the vessels and into the heart, shoot radiopaque dye into it, and take moving X rays of that dye as it passes through all parts of the organ. The technology is so precise, they can shoot dye into the little coronaries and see exactly which ones are blocked and need to be bypassed. Such techniques can also be used to *fix* certain problems. A cardiologist can insert a tube through the vein in a patient's leg, push it on into the heart, then blow up a tiny balloon in a constricted vessel (angioplasty), or place a small brace called a stent in there to ensure it will stay open, or even plug up a hole between the atria.

Here, then, was an entire medical specialty that rested on the work of doctors who could reasonably be called crazy. Imagine the questions people must have asked before the first open-heart surgery: *You want to cut open this child's heart and fix it inside?* How? *You want to hook this child up to another person and have that person's heart pump blood through both their bodies, to supply oxygen to them while you cut open the sick child's heart?* You *try it;* I'm *not gonna try it.*

Called cross circulation, this technique was developed by Walt Lillehei in 1954, and it worked more times than not in forty-five operations. This was the way cardiac surgery began, with big, dangerous experiments. In 1952, even before coming up with his cross-circulation idea, Lillehei had assisted John Lewis in closing an atrial septal defect simply by cutting off part of the patient's circulation; the following year John Gibbon Jr. became the first to use cardiopulmonary bypass in the same procedure, ushering in the modern era of heart surgery. These surgeons have rightly been branded mavericks. It's easy enough to say, Well, in most of these cases the patient would have died anyway *without* their surgical intervention, so what did they have to lose? But the fact remains that the medical profession is founded on a single tenet: First, do no harm. In Lillehei's cross-circulation case, he risked killing his patient simply by opening up her chest and cutting open her heart (he had killed others before her), but what really clouded the matter—and prompted others to dismiss it as madness—was that this was by all accounts the first surgery ever performed on a single patient that had the potential of resulting in two deaths, the patient's and the circulating donor's.*

And who, I wanted to know, first discovered that a tube could be threaded into a person's arm or leg and wind up in his or her heart without *killing* him or her?

His name, it turns out, was Werner Forsmann, and he was a German doctor. There inevitably has to be a human guinea pig in such experiments, as there was when Walt Lillehei decided to attempt the first-ever cross-circulatory VSD closure. Forsmann's guinea pig was himself, and he performed the procedure in secret, having been forbidden to attempt any such thing by his boss (first, do no harm to yourself). Forsmann thus became, in 1929, the pioneer interventional cardiologist by feeding a tube into a vessel in his forearm and pushing it through his body till he felt it reach his heart.

Growth in the interventional cardiology specialty "is occurring at a snowballing pace," says Larry Latson, the head of cardiology at the Cleveland Clinic's congenital heart center. As the procedures become safer, he adds, as devices get better and smaller, as more innovations such as valves delivered by catheter are developed, and as catheterization itself is more fully integrated into the long-term management plan for each patient, the rate of that growth will become harder and harder to predict. The clinic does, on average,

---

*The development of this operation and the origins of open-heart surgery are compellingly described in G. Wayne Miller's book about Lillehei, *King of Hearts: The True Story of the Maverick Who Pioneered Open-Heart Surgery* (New York: Times Books, 2000).

9,000 heart catheterizations, 6,500 diagnostic procedures, and 2,500 interventional caths each year, 350 to 500 of them on children. In 1996, the total number of catheterizations performed nationwide was 472,000, according to the American Heart Association's 2002 statistical update; by 1999 the total had jumped to 1.3 million. And new techniques are continually being introduced: in September 2001, for example, a team at Brigham and Women's Hospital in Boston opened up an underdeveloped aortic valve in a fetus—a potentially deadly problem in a newborn—in the first-ever successful in utero repair of such a dangerous heart defect.

There is today no small ego battle being waged—most of it tongue in cheek—between interventional cardiologists and cardiac surgeons, as their purviews merge. Interventional cardiologists are the thinking man's surgeons, insist the interventional cardiologists, doing the same work as the surgeons with none of the mess or long healing time. The surgeons, for their part, roll their eyes and look pityingly at the interventional cardiologists—wanna-be surgeons all, they intone. Interventional cardiology is elegant on the outside and ugly on the inside, Mee has often said. Surgery is ugly on the outside but elegant on the inside.

This whole world, unknown to most everyone, astonished me. And the fact that the ultimate reason for this specialty was to save the lives of babies and children endowed it with a brutal power not evoked by the treatment of adults with acquired heart disease—patients who had in most cases already lived long and productive lives and whose problems were the result, often, of inactivity, the consumption of too much fat and too many cigarettes, or perhaps simply being alive and well for seven or eight decades. In peds the patients were typically infants who had done nothing more self-destructive in their short lifetimes than be born, and the surgeon who took one of them into the O.R. usually had the weight of an entire family on his shoulders—all the parents' hopes and dreams for the imagined future of their baby. That the surgery itself was never easy only upped the ante even higher. This work required real hand skill; it wasn't just a matter of "laying pipe," as standard by-pass grafting is sometimes referred to. The peds surgeon needed a heightened ability to turn flat objects into three-dimensional ones in his mind and then repeat the process with scissors, needle, and thread. He (it is an almost exclusively male domain) had to be a wizard at patching and sewing; he had to be fast; he had to know where all those invisible conduction systems were buried in the heart muscle; he had to have an intuitive sense of what kind of blood pressure was needed in those vessels and what kind of volume these delicate, elastic vessels could handle; he had to be capable of thinking on his feet once he was in there, because no heart is ever like any other; and when something went wrong, as it did all the time, he had to be able to increase the volume of his calm.

The notion that this type of surgery was particularly difficult led me to another idea: If it was that hard, I reasoned, some surgeons must be better than others. Obviously, as in any profession, there must be a range from worst to best, with most falling somewhere in the middle. But here, where the stakes were so high, how could anything short of the best be tolerated? How could a surgeon be on the low end of the scale and still be allowed to operate? A sloppy accountant was one thing, or a lazy carpenter, or somebody just OK at most any job—but an inferior children's heart surgeon? It was a frightening thought. Yet there was no oversight at all over the peds centers, except where the surgical outcomes were completely unacceptable, or where pressure from the media and outraged parents, perhaps sparked by a whistle-blower in the hospital, prompted a widespread investigation. In the year I was at the clinic, scandals would erupt over the quality of surgery at children's hospitals in Denver and Seattle, and the Bristol Royal Infirmary would close the books on a major investigation into surgeons who had continued to operate on babies despite having intolerable mortality rates. The stakes were high here; in fact, there were none higher— the kids' lives hung in the balance. The aspiring pediatric heart surgeon had to be great from the get-go—but how could anyone be excellent, let alone perfect, the first time out? This specialty warranted scrutiny, and it no longer had room for a learning curve. Yet kids with fixable problems were still dying, and there were still good surgeons and not-so-good ones—the same range found in every profession.

Given all that, how did someone get to be a peds heart surgeon?

I wondered. Complex cases could not be handled exclusively by the best surgeons; if they were, no one else would know how to deal with similar situations after those surgeons retired or died. There had to be a way of replenishing the system: young surgeons would have to do their first arterial switches, their first Norwoods, at some point, or eventually no one would know how. So there had to be firsts, absolutely—*but not with my kid*. So if not with mine, or with yours, then with whose? Should the novices practice on pro bono cases from Ecuador, or maybe on Medicaid cases from rural Kentucky? It was wonderful when a child lived because of what a surgical team did, or when a six-year-old born with hypoplastic left heart syndrome, once uniformly fatal, came to the hospital to say hi to everyone, looking for all the world untouched and in perfect health; but it was almost unbearably difficult for the parents and children and doctors and nurses and technicians when things went less well, as they frequently did. Thanks to Mee's reputation, his center tended to get sent the worst cases, the stuff no one else wanted to touch, such as repairs that had gone bad elsewhere and were now considered hopeless. In more than one instance parents had been told by their doctors, "We don't believe anything more can be done for your child, but we can write to Roger Mee in Cleveland to ask if he thinks *he* can do something," and often Mee had replied that maybe he *could* help, even if he had no real way of knowing one way or the other, only a hope, a guess. In the case of an infant boy named Drew H., for example, he told me, "I just don't know if I'm going to assassinate this kid." But because the only other option for Drew was a fairly quick death, he was willing to risk it—as were the child's parents.

There were, I would gradually find out, only a handful of surgeons whose reputations rivaled Mee's. Regularly mentioned within this small community were Ed Bove, who practiced in Michigan, just three hours from Cleveland; Frank Hanley, the youngest of the marquee peds heart surgeons, then at the University of California in San Francisco; and the team at Children's Hospital of Philadelphia. In England the standouts were Marc de Leval at London's Great Ormond Street Hospital for Sick Children and

Bill Brawn in Birmingham; the Hospital for Sick Children in Toronto had Bill Williams; and in Paris there was Claude Planché at Marie Lannelongue. The cardiac unit at Children's Hospital in Boston, headed by Richard Jonas, was always cited collectively as the standard-bearer: it was at Children's that the very first closed-heart surgery on a child had been performed in 1938, there that Aldo Castañeda had initiated the move toward neonatal repair in the 1970s. It remained the biggest and arguably the best in the country in terms of facilities, overall care, and number of patients treated (in this specialty, this last item was apparently directly related to the quality of care offered).

The more I learned about congenital heart surgery, the more I wanted to explore the surgeon's mind. It could reasonably be argued that this was the most difficult surgical specialty in medicine, given the technical complexity of the work itself, the age and weight of the average patient, the infinite variety of defects involved, and the emotional burden of operating on neonates, infants, and children. Plastic surgery demanded extraordinary skill, and those surgeons who could return a burn victim to some semblance of human appearance deserved honor for their efforts; but plastic surgeons didn't by their skill alone determine whether their patients lived or died. Neurosurgeons seemed high up on the mystique totem pole, operating as they did within the organ of intelligence, often viewing their field through scopes of enormous magnification, and, like peds heart surgeons, working in hair's breadths, with the risk of permanent, debilitating damage on either side of that hair. Neurosurgery was deserving of its mystique, a specialty requiring serious skill, but in the end that skill was largely a matter of sucking. Brain surgeons, for their part, mainly take stuff out. In fact, the majority of *all* surgery was the removal of unwanted tissues, was destructive by its very nature. In *their* work, by contrast, heart surgeons repaired and rebuilt. All surgeons, those men and women who operated inside the human body, had a certain romantic hold on the imagination of the laity, but within the medical profession, I knew, they were considered the lugheads, the plumbers and mechanics; physicians liked to think of themselves as the *brains* of the profession, but it was clear who had more clout: the physician was the straight-A student, and the surgeon was the scholar-athlete.

The peds heart surgeon ("peds" with a long *e* as in "pediatric") had to have it all—the intellect and the skill, the grace and decisiveness, the sewing skill of the craniofacial plastic surgeon and the vascular surgeon, the millimetric precision of the neurosurgeon—and he had to be fast, too, because the longer a kid was on bypass, the longer a baby's circulation was completely shut down, the harder recovery would be, and the more danger there was of what was euphemistically called neurological deficit. There was no other surgery in which precision combined with speed mattered so much. Add to that the need for imagination, and the bar began to seem almost insurmountably high.

The peds heart surgeon had to be perfect because the heart didn't tolerate imperfection, didn't right itself, didn't compensate. "Close enough" didn't work here. The peds heart surgeon often had to build new structures within hearts that measured little more than an inch or two in any dimension. He had to be a quick thinker, too, because often what had been diagnosed wasn't exactly what he found on seeing the actual heart; he had to make judgments—about the long-term strength of a ventricle, or how big a shunt to use to send how much blood to the lungs—he had to make decisions fast and act decisively, and he had to be right. (The old chestnut about surgeons, "Sometimes wrong, never uncertain," didn't seem so comforting in peds heart surgery.) And he had to do it all knowing that a mom and dad, and maybe grandparents and siblings, were pacing in the parents' lounge, hanging on his skill. A slew of techs, nurses, perfusionists, anesthesiologists, and doctors was responsible for each child, but it was the surgeon who had the greatest effect on the final outcome, for better or worse; he had to recognize that his skill and intelligence would determine how good or bad this child's life would be after his or her chest was sewn back up. Open-heart surgery on kids was a complex team effort, but the surgeon was the one who would bear most of the responsibility, and the resulting glory or shame.

Granted, the world of congenital heart surgery was small and select, with negligible impact on the national health scene and little part in the dominant debates on public health, but I couldn't imagine a more intriguing, sometimes astonishing, often harrowing, and inevitably dramatic medical specialty.

"It's the top of the shit pile," Frank Moga, a surgical fellow at the clinic, says of peds heart surgery. Frank is thirty-five but looks ten years younger, a wiry cur with dark, wavy hair and eyes that always seem to have that postcall wired dementia, as if they were seeing great comedy or great horror, it's impossible to say which. It was Frank who once described the atria for me as "nice distensible bags" and the ventricles as "beautiful muscles." He loves the heart; to him it is a gorgeous structure, the heart of the child—lovely to behold in its shape and its rich and amazing colors, which run from yellow and pale pink to bright red, purple, and blue—and operating on it, in his estimation, approaches the sublime. "In medicine, it's the closest thing to art there is," he says.

Peds heart surgery is among the hardest and most prestigious medical specialties because it's so difficult to do well, because of what's at stake, and because of the complexity and variety of the physiologies involved. It takes more years of training than virtually any other specialty, and even someone who gets through that training and manages to begin his actual practice by age thirty-five or thirty-six can't count on being a success. Moga says that some who attempt it simply don't make it in the end, either because they aren't good enough or because the risk of failure eventually scares them off. "First couple of kids you kill," he explains matter-of-factly, "the cardiologists stop referring you patients, and that's it."

Where Roger Mee fitted into this picture, I couldn't yet be sure. I knew what his reputation was, and knew, too, that he had likely seen more congenital heart defects than just about anyone else currently practicing. He was accorded respect by colleagues all over the world for his experience and infinitesimal mortality rate. His secretary and department coordinator, Debbie Gilchrist, would later tell me about the time she accompanied him to a huge congenital heart conference. Gilchrist is a thirty-something bundle of effectiveness and can-do energy, with big red hair and a mouth that speaks her mind. ("Living in this world will *change* you," she confides. "It makes me value life. When a friend has a healthy baby, I don't think it's normal; I think it's a fucking miracle.") She saw at this conference how others in the field acted around her boss, and it amazed her: strangers hovering in hopes of getting a word with him, a glimpse of him, or even reaching out and touching him, as if he were some kind of deity.

"This short bald guy who scratches his belly!" she would say with delight.

Mee was not universally adored, to say the least. No one doubted his ability, but he was rumored to be hell on his residents, and a member of the clinic's board of governors would inform me that he was considered "difficult." And a former clinic surgeon would allow of Mee, "The guy can cut. He's a great technician, no question. But he's a tyrant. For him, mistakes don't exist. He's cold with parents. . . . But hey, when you're Michael Jordan, you can do what you want."

I spoke with Richard Van Praagh in Boston, one of the éminences grises of this specialty, a cardiologist and pathologist who has classified many different forms of heart disease. He remembered Mee fondly from his six-month rotation there and said, "Roger Mee is one of the very best heart surgeons in the world," adding with a jovial chuckle, "He's not an innovator. He's known for doing everyone else's operation better than they do it themselves."

Mee was a virtuoso within a small and elite surgical specialty, whether by accident or design, a stroke of luck, or by some other grace in this peculiar world. He was like an unexplainable sports phenomenon, a Jordan or a Larry Bird, or like Glenn Gould with Bach's *Goldberg Variations*. It seemed to have nothing to do with *him:* he didn't love to operate, he said, and whereas some surgeons lived to be in the O.R., he eagerly awaited retirement. But he had been touched, like any of those figures who describe genius through their bodies' moves and work. And so he did that work.

As Dr. Floyd Loop, a cardiac surgeon himself and clinic CEO, would tell me, Mee's results were second to no one's, and that he got such results out of the very worst cases meant something. Mee was regularly listed among the top few surgeons practicing. More-

over, he had reached that pinnacle by doing it his own way. His repairs were idiosyncratic; at conferences, surgeons presenting papers
often used his name as a virtual synonym for an alternate method.

In one presentation I would see, however, Mee's achievements
would be conspicuous by their absence. A prominent surgeon cited
his and other top centers' results as proof that switch operations
could now routinely yield, as his center's did, a 3 percent or lower
mortality. Afterward Mee called out to this surgeon, a friend, "Why
didn't you ring me for my results?" The surgeon didn't respond—
and didn't need to. Mee's results—less than 0.4 percent mortality—
would have made the best of the rest look bad by comparison. He
was an anomaly.

When Mee was a boy, he didn't much care for school, preferring rugby, gymnastics, and boxing to books. He was spontaneous,
disappearing on sudden trips whenever the mood struck him, even
if it meant missing a date with his future wife. He went into medicine for lack of any better alternative. Raised by professionally religious parents, he was religious himself and figured he'd end up as a
missionary physician in Africa. The graduating medical student
wandered into cardiac surgery more to avoid becoming bored than
because he felt any particular affinity for it. He became a pediatric
heart surgeon, too, by chance, after an older colleague recruited
him to run a problematic congenital heart center in Australia. Mee
was just finishing up his residency in cardiac surgery at the time,
and he had done a six-month rotation in congenital, so he accepted
the job. Within five years his center was reporting better results
than any other in the world; Mee was still in his thirties.

In his prime, he told me—in the 1980s and early 1990s—he had
performed more than seven hundred surgeries in a year, a stagger-
ing number ("If you're doing more than two hundred and fifty operations," one surgeon explained, "you're in the O.R. *all* the time").
Now in his midfifties, Mee was nearing the end of his career, and
his annual quota of three hundred to four hundred surgeries was
getting to be too much for him; he would have no difficulty retiring, he said, just as soon as his youngest child was out of college.
Earlier he had also conceded, "There comes a time when your increasing wisdom and experience do not compensate for your de-

creasing ability in manual skill," a variation on one of the fundamental dictums he'd repeat to me on several subsequent occasions:
"A surgeon has to know his limitations."

Mee didn't fit any mold or stereotype. He wasn't manic or lunatic, like some other surgeons, and what was more, he didn't seem
to have a colossal ego. (When I mentioned this latter observation to
a physician friend, he exclaimed, "That's the worst kind!" Then he
told me a joke: How many surgeons does it take to screw in a light-
bulb? One—he holds the bulb while the world revolves around
him.) Granted, there must be an ego *somewhere* in there, but whatever form it took, Mee's wasn't an obnoxious ego, or one that he
used to belittle people. Several of his colleagues would remark how
smart Mee was or tell me that he was the smartest *surgeon* they'd
ever encountered (a backhanded compliment). But he was by no
means an intellectual, and even denigrated intellectualism. He was
articulate but not a pontificator. He didn't like the convoluted
what-if discussions of ethics that physicians thrived on—"wanking," he called it. Mee seemed not the least reflective about what he
did or why; he was just "a regular bloke," he claimed, a fan of hot
curries and single-malt whiskey. Cardiac surgery suited him, he
said, because the heart provided immediate feedback: if he got a
procedure right, the patient's heart would come back instantaneously, and if he got it wrong, it would stop beating properly.
Heart surgery was *clean*—meaning it wasn't like gut surgery, no
messing around with feces or pus. Repairing a young heart demanded greater precision than sewing up a gut, and Mee seemed to
like that requirement, seemed to enjoy having to be perfect.

"It's clever, but it's primitive," he would later say of the organ
that is his milieu. "It doesn't do much. Its primary function is to
generate energy. The liver is more complex. The kidney is more delicate. The brain is"—and here he paused, a look of reverence on his
face—"the brain is *exquisite*." I asked him if he liked the heart, and
he chuckled and grimaced at me as though it would take some
doing to top that one for stupid questions. As I recall, he was in his
shirt and Skivvies at the time, in the process of changing into his
white O.R. clothes—something he did right there in his office,

with the door open (if June, the cardiac nurse, walked in, X rays and cards in her hands, she'd throw back her head, turn on her heel, and leave the room till he got his trousers on). "Do I *like* it?" he repeated to me, balancing on one leg. "I *hate* it. It dictates my life. It's my ball and chain." (On other occasions, trying to draw him out and finding it difficult, I'd ask him how he was "feeling"; he'd parrot my question, making new-agey fun of me but not answering.)

There were all kinds of fascinating things going on in medicine that would have made other specialties or other centers great subjects for me. Other medical fields had bigger public-health implications, certainly, and a trauma unit in New York City, for example, would no doubt have been more dramatic, more gruesome. But the ingenuity required to fix a baby's heart—the puzzle of it—enthralled me; the advances in pediatric heart surgery were still new; and it was still a relatively young specialty, little known outside its own confines. The drama of the O.R. was real and intense. The surgery itself was a genuine craft, and here I'd met one of its foremost practitioners, a craftsman who might even be said to attain, as some have claimed, the level of art. And art in this case might be defined as not just life-affirming but actually lifesaving. In Mee's hands, art meant a two-ventricle heart instead of a one-ventricle heart, or in some cases, a child's leaving the hospital a week after having major surgery rather than lingering for a month on a ventilator and being incapacitated ever after.

Mee made it clear to me that he didn't think much of lay writers who wrote about medicine. But he was independent enough personally, and within the clinic, he was sufficiently mollified by my focus in previous books on excellence in craft (that word *excellence* especially intrigued him); he hoped I'd concentrate this time out on the whole team, the big picture, he said; and he evidently liked my manner enough to tell me, ultimately, that if I wanted to observe his center, I was welcome to do so. His only stipulation was a reasonable one about honoring patient privacy; otherwise I'd have no conditions placed upon me. Deb would take care of getting me scrubs.

3. Bad Hearts

Angie H. shakes her baby, scared, desperate to quiet him. This is her first morning in room 1 of the Cleveland Clinic's Pediatric Intensive Care Unit, and her four-month-old son, Drew, is in such distress that he's issuing weak, awful cries and twisting in her arms. He's been like this for most of the night, ever since he arrived by ambulance from another hospital—in fact, he's been like this for two days now. Shaking him in her arms—it's not a rocking or a gentle bouncing but instead a vigorous agitation—seems to be the only way Angie can distract him from the discomfort of his failing heart and starving tissues.

Angie is a twenty-five-year-old mom with blond hair and bright-blue eyes; she's pretty but visibly exhausted and afraid for Drew, her first child. Her husband, Bart, stands behind her, tall and solidly built, fair-haired and blue-eyed like his wife. He's quiet and expressionless when the team of surgeons, cardiologists, intensivists, and nurses enters the room on morning rounds.

Drew is pale, with legs so skinny that Angie and Bart called him Chicken Legs during more hopeful times at home. Drew writhes in Angie's arms. Angie shakes him. I saw her shake him earlier, when I passed by the room, and will see her do it throughout the morning. It's uncomfortable to watch, not only for all the obvious reasons but also because she has to jiggle him so hard, and he is squirming so furiously to escape her hold, twisting his torso and crying, that I keep thinking she'll drop him. It's a nerve-racking sight. He's actively trying to go crashing to the floor, and Angie, while shaking him in her arms, is hiking up her hips under him, one at a time, to prevent him from falling.

It's not that he's in direct pain; rather, his heart muscle is dying, scarcely able to pump blood to his body, and his tissues are screaming for nourishing blood. It must be the kind of distress you'd feel if there weren't enough oxygen in the room and you kept taking deep breaths but could never get enough.

Drew's birth at a community hospital the previous July went smoothly, but twenty-four hours later he began to turn blue and was diagnosed with transposition of the great arteries. He was transferred to a Pennsylvania hospital, where several days later he underwent a switch operation. On opening his chest, the surgeon found an intramural coronary—the big risk factor—and the complex switch was completed with bad results: the repair had the unintended effect of cutting off the blood flow in the left coronary artery, which feeds the left ventricle. Drew eventually recovered enough to go home, and his doctors hoped that his left ventricle would in time regain normal function, either by itself or through the development of collateral arteries from the right side of the heart. But it didn't. Drew's cardiologist subsequently wrote to ask Mee for advice; Mee said he thought he could help; and the family made plans to travel to Cleveland. The night before they were to leave, however, Drew was admitted to the emergency room at that same Pennsylvania hospital, in severe distress. There, doctors filled him with inotropes to help his heart beat, and the following day he was taken by ambulance through the snow to the clinic, several hours away.

The clinic doctors discussed Drew's case yesterday, in their regular Monday-morning catheterization conference, along with the cases of the numerous other patients who are scheduled for surgery this week. By midday, though, word began to circulate in the unit that the baby was being rushed in due to myocardial ischemia, or lack of blood to the heart, and his surgery had been moved up to the next day—today, Tuesday, November 20. Rounds are quick in room 1 this morning, beginning with a brief history of Andrew H.:

hospital transfer, four-month-old boy, status post–arterial switch operation, scheduled for surgery this afternoon to try to fix one blocked coronary. An attending PICU doctor, Steve Davis, reiterates the powerful inotrope dobutamine that should be administered to help the heart's pumping. Dr. Mee stopped in to see Angie last night after Drew's admission, and he asks her now, "Did you manage to talk to him?" Bart had to drive through the snowstorm in their truck and was late getting here. Mee wants to be sure that both parents understand how serious the situation is and what's going to happen later today.

"I tried," she answers, a gentle lilt to her voice. "I didn't explain it as well as you could."

Mee nods and tells the couple, "I'll come back to speak with you if you have any questions."

Steve Davis was the one who admitted Drew yesterday, and he says privately to surgical fellow Frank Moga, "I'm having trouble convincing people that he's the sickest kid here"—mostly because he isn't intubated and has enough energy at least to cry and writhe. Drew's ejection fraction is deathly low—the left ventricle is pushing out only 8 percent of the blood it contains—and the blood from the rest of his body is thus backing up in his lungs and stagnating everywhere else. An initial echo showed that the left ventricle is scarcely moving. Steve called Roger at home to give him a status report on Drew, and Roger asked him to put the kid on a ventilator. "Roger wanted him intubated," Steve tells Frank. "I did not want to do that. I've seen three kids die from decreased cardiac function." Frank nods and grins: *Right, I know.* Steve had to call Emad Mossad, the head anesthesiologist, for support, and they both talked to Roger, who at last "convinced" Steve that it would be safer to wait and intubate Drew today, when he's in the O.R. *"Thank you,"* Steve says, his eyes wide. Frank, laughing, sticks his knuckle in his nose and makes a face. I ask Steve why Roger didn't appreciate the danger of intubating the baby, and in unison Frank and Steve say, "Because he's never had to do it before!"

Drew will be watched carefully all morning, until the O.R. staff arrives to take him to surgery. The doctors leave the room to round on the next of several more kids with terrible hearts: hypoplastic left heart syndrome, pulmonary atresia, a nine-year-old girl recovering from her second transplant, a baby born with his aorta and pulmonary artery melded together who has produced not one drop of urine since his operation a month ago. To call Drew H. the sickest kid in the unit is a serious claim.

Cardiac rounds are the morning constitutional of the unit. Night turns to day when the nursing staff changes over, and at seven A.M., a fresh shift clusters behind the main desk to get a description of each ICU patient's condition. The day's attending intensivist rounds on each patient, either alone or with the previous night's attending. (This PICU prides itself on always having an attending PICU doc in the building, rather than assigning night and weekend call to residents who are still in training. The routine was established shortly after Mee's arrival, when the clinic began attracting very sick cardiac patients who required around-the-clock expertise.) And then at eight A.M. a scrum of one or two dozen docs, nurses, therapists, social workers, pharmacists, and visiting surgeons fills the corridor and moves with glacial speed down its entire length, Roger Mee the focus of the resident doctors who present each cardiac patient to the group.

It is during rounds, the daily updating of each patient's status, that consensus is reached on a plan of action for every case. This is also the only time during the day when each type of caregiver is represented in the decision-making process. Rounds can be the scene of abstruse debate, the dressing-down of a shaky new resident, or routine maintenance, but when the unit is filled with complicated hearts and sick kids, as it is now, the scrum seems to me collectively like a chess pro moving from game to game, assessing the board, making a move, then stepping on to the next board. Sometimes rounds can also be, as Steve Davis puts it, "an endless spewing of numbers," with the real care being provided by the ICU docs and cardiac nurses and other ICU staff once the crowd clears.

"By the end of rounds, people are numb," Steve tells his residents. "Try throwing in a crazy number at the end of rounds, and I guarantee you no one will notice except maybe Cheryl or June." Cheryl and June, the two most senior cardiac nurses, are responsible for knowing virtually everything about each patient's condition and recording it all on four-by-six-inch note cards.

Steve grew up in a working-class neighborhood outside Boston and slogged in his family's Italian restaurant as a youngster. His height, five-eight, he says, precluded any realistic chance of his having a professional baseball career, so instead he made his own way through Johns Hopkins, then med school at the University of Vermont. Sturdily built, with dark hair that's cut close, a can of Diet Coke usually in hand, Steve prefers jeans and a sweatshirt to the white clinician's coat or the shirt and tie that seem to be the norm among the intensivists here. He gets impatient during rounds, wishing the docs simply had a stamp for the routine cases. Even Mee, noting how quickly the scrum sometimes covers the ICU, acknowledges to Steve, "ASDs, you don't even notice them." The ASD, or atrial septal defect, a hole between the atria, is a common problem, and one that can easily be repaired, with no expected mortality. If the hole is small enough, an interventional cardiologist can even plug it up in the cath lab, so the chest never has to be opened at all. Here, where so many bad hearts need fixing, no one even blinks at an ASD anymore.

Steve tends to speak in sports metaphors, calling himself and the other intensivists in this unit the team's "utility infielders." When his backup, Kathy Weise, arrives one morning a few minutes late, he describes the entire unit to her this way: "No one's actively dying." Turning to me, he says, "No one dies quickly here. We don't let anyone die quickly. They all die slow, horrible deaths. We tackle them on the one-yard line. And we've got a *great* goal-line defense. They can be *sprinting* for the end zone, and we get 'em."

This fourteen-bed PICU admits about a thousand kids a year, more than half of them heart patients. On average, fifteen of the thousand don't leave alive, and of those, ten are cardiac patients (an overall 1.5 percent ICU mortality, with cardiac mortality "consistently under two percent," according to Michael McHugh, the head intensivist). The typical stay is between one and two days, or three to four days for cardiac patients. About 20 percent of the five hun-

dred noncardiac patients are neurology cases who have had operations for tumors or seizures. This PICU is considered medium-size,
with six beds being considered small and more than fifteen big.

Steve Davis says that with cardiac patients, much of his and his
colleagues' success is the result of excellent surgical repair. "We can
fiddle with the ventilators all we want," he says, "but if there's a
shitty repair, there's nothing we can do. I've been in places where
they come out of the O.R., you *know* they're going to die in two
days. And in two days, they die. These guys are so good that we're
able to keep them alive. Brandon Deibler should be dead. He's not
out of the woods yet, but . . ."

Brandon was born in September with hypoplastic left heart
syndrome, in which the left ventricle is more or less nonexistent.
An HLHS diagnosis used to mean certain death, but now there's a
procedure called a Norwood that can save many infants with this
defect. Nevertheless, the Norwood still has a high mortality rate,
even at the best centers. Brandon didn't exhibit his condition until
he was two days old, at which point a surprised nurse noticed how
mottled his skin was and that he wasn't eating. Within hours he was
in Cleveland, where Jonathan Drummond-Webb did a Norwood
on him. The postoperative course has gone badly: Brandon has
been unable to be weaned from the ventilator, and he developed a
chest infection. If anything *can* go wrong with him, it seems to do
so. Immediately after a postop catheterization procedure was performed, he arrested; a code was called, and he was resuscitated. No
one expected him to live after that near-catastrophic insult, nor
could anyone say what kind of cerebral damage might have occurred. Serious internal bleeding, most likely resulting from the
cath, began to fill up his belly with blood; Jonathan had to take
Brandon back to the O.R., cut him open again, and dig around in
his bowels to search for the source, a tangle of intestine rising up
around his wrists as he rooted. Still the bleeding continued, until
Jonathan took him back the next day and fixed the problem. Then
Brandon became septic; his blood was infected. Steve Davis looked
at the mom and dad, Melanie and Tod Deibler, and said, Melanie recalls, " 'I know what you want to ask me. You want to ask if your
baby's going to live. I don't know.' "

"That kid really took a hit," Mee told me after the catheterization incident. "If he comes through, it'll be a real tribute to critical care." He regretted, he said, that no one would admit to having made a mistake—the likely scenario here. "It's always the kid's fault or something they can't figure out," Mee said. People had to be more honest with themselves, he believed, though he knew how hard it was for any doctor to say, "I went in there to try to help this kid, and I hurt him"—particularly in a culture that seemed to prefer pointing fingers to putting any effort into prevention. "Americans love to blame," noted this New Zealander.

But Brandon has pulled through even this last episode. Each day during rounds, variations of surprise are murmured. Jonathan says, "It's amazing how resilient he is." Moga agrees: "He's a tough little kid." And then, amazingly, "Can we think about extubation?"—a small landmark in the recovery process, especially for kids such as Brandon, who have virtually never breathed on their own. (Steve Davis is convinced that if the failure rate of extubations in a PICU is under 10 percent, it means the docs aren't being aggressive enough in getting patients off the ventilator—a misguided caution that can ultimately result in lung disease and other problems.) The doctors are even optimistically thinking that if they can extubate him, Brandon may soon be ready to be moved to the "floor," or the step-down unit, out of the ICU. When Jonathan asks Steve if he'll be able to bear it, Steve hesitates and shakes his head: "Separation anxiety," he says. He considers the situation, then suggests, "Maybe when he's in kindergarten. We'll send him directly to the bus: 'That's OK, Brandon. We know the other kids don't have central lines. They won't make fun.'"

Across the hall from Brandon, another case is having a less happy result. Leaning on the doorjamb as the scrum files in, Marc Harrison says, "He's a frustrating kid to take care of."

Luke was born prematurely with his aorta and pulmonary artery welded together. Before surgery became routine, this defect was fatal more than half the time, and where it wasn't fatal, the prognosis was grim. Beginning in the early 1980s, though, neonatal repair could prevent the terrible lung disease that was often a sequela for kids with the repair. Luke happened to have an obstructed aorta as well, which made this procedure particularly tricky: Mee put the risk of death or near-death at 10 percent, an unusually high number for him. He confessed before the operation to being nervous about this one and said he had tried to plan it out in his head the previous night. Not only did he have to separate the pulmonary artery from the aorta, put in a new pulmonary valve, and close the VSD, he also had to rebuild the aorta itself—all in the very small heart of a very small baby. But when the operation was finished, he knew he'd done a good job. "The repair was perfect," he said.

Which makes Luke's moribund state especially frustrating. The team has been scratching its collective head each morning for weeks on end in his room, which looks like a small technology center by now, with a battalion of medications being administered through electric pumps that hang beside the warmer bed, the ventilator, the monitors, and a dialysis setup. "The more crap you see in the room," Frank tells me, "the worse it is."

After all the numbers are read off by the resident, everyone stands and stares. Adel Younoszai, a young cardiologist, asks if they should consider doing a catheterization to get better hemo-dynamic data.

Mee says no—if you're not looking for something specific, it's not a good idea to do an invasive procedure on a critically ill child. A colleague adds, "It'll be hard to give him dye if he's in renal failure." Adel nods. Roger asks if any of the residents wants a project: find out which anuric infants have survived. Steve says he's never seen a kid not come back after renal failure here. Frank Moga suggests trying vasopressin to tighten Luke's blood vessels—"really clamp him down," he says. "The right pulmonary artery is right on top of the aorta, and that may be hindering blood flow. I think it's affecting cardiac output." They try it, but nothing happens. Are the kidneys dead? someone wonders. If so, they should have shrunk, and they haven't. Besides, an echo technician has confirmed that there's blood flow to the kidneys. So what's going on inside this baby?

Luke has lain unconscious in the warmer bed for so long that nutrition has become a serious issue, because only so many fluids and drugs can be put into a kid who's not putting anything out. Trying to address the problem, Marc Harrison inserted a Corpac, a special feeding tube, but he pushed it a fraction too far and perforated Luke's duodenum. Because the baby is on anticoagulants, he bled and had to be transfused. Beyond the obvious damage, Marc also worried about infection. During the next day's rounds, he announced his error and accepted the blame for it. Roger asked, "Will this do him in?" "No," Marc answered. Indeed, the insensate organism that is this child's body has proved surprisingly eager to go on living. Still, Luke's parents, who rarely visit him, are asked to come in and discuss a DNR ("Do Not Resuscitate") order, and they agree.

Mee stops in to check on the baby boy when he walks through the PICU. In rounds there was talk of putting in a new arterial line, since the old ones kept getting infected, but no new line has been placed. Mee asks a nurse, "Has Steve given up on him?" Steve himself has meanwhile entered the room, and he answers, "I haven't given up on him. I don't want to be active in his death." The kid's got platelets of 20,000, he reminds Roger, and he's on heparin, an anticoagulant; his tissues are edematous, sodden with fluid, making access to vessels difficult and the possibility of fatal error higher than he'd like. He says he'll wait for Frank—"I want Frank to spread out the responsibility when he dies."

Luke is simply a regular part of rounds, just like Ashley Hohman and Julia Stone and Brandon Deibler, all of them unconscious and on ventilators, three of the four of them never having breathed on their own. The situation seems so unhappy, one morning in Luke's room, Roger asks, "Can we save the heart and do a body transplant?"

This prompts laughter as well as some more serious references to transplantation, oblique at first until Jonathan says, "I think if we can find a heart for Ashley Hohman, we'll have the kidneys." Maryanne Kichuk, the cardiologist who heads the transplant program, adds, "They'd be thrown out anyway"—thrown out because kidneys that size, kidneys from a two-kilogram baby, have never before been successfully transplanted. The idea clearly intrigues

Mee. "I don't see why it couldn't be done," he says. "And we should do it"—"we" meaning pediatric cardiac surgeons—because "we're very good with microvasculature, and we're quick."

Steve listens as the notion is debated. He turns away and shakes his head until finally he can't stand it anymore. "It can't be done," he says. There are young doctors present, residents, and Steve, who's forty-one, thinks the ridiculousness of this discussion needs to be pointed out to them. When the conversation resumes, he says again, more forcefully this time, "It can't be *done*. I don't know why we're even *discussing* it."

Jonathan snaps, "They never did a switch before, so to me that's a nonanswer."

"The smallest they've agreed to do is six kilograms, and we shipped her off to Minnesota. They've been keeping her alive for eight months now."

The exchange continues in spite of Steve's rolling eyes. Jonathan asks Maryanne, "Should I talk to Cunningham?" Cunningham is the head nephrologist.

Steve actually throws up his hands and leaves the room, saying, "I gotta start eating breakfast with you guys."

Marc says, "He's only on eighty percent maintenance. On eighty percent, he won't grow."

Mee nods and says, "We've got to get a little more into him."

In the end nothing comes of all this, nor does a heart appear for Ashley. "They were blowin' smoke," Frank says after rounds. Steve is irritated with Marc for even letting them talk about it: "Maybe you're going to transplant a kidney in a two-kilo baby in five years, but it's not going to be in *that* one"—a baby with a bad heart who's been on a ventilator for a month. Marc nods. Steve tells me later, "It was a lot of posturing. But it's typical: Roger and Jonathan's response in a situation that's desperate is they never want to give up. They take it personally. I don't. It's not about me. But I've never had my hand inside a kid's chest, working on a heart. I think that might create some kind of bond that makes it harder for them.

"I haven't given up on him," he says of Luke. "But like that situation the other night with the central line: I don't want to do anything that's gonna harm him, in that case maybe kill him. Now,

Roger's in the operating room every day. Every day he does things that could possibly kill. But there, there's a lot of potential benefit. Here I didn't see a lot of benefit."

A week later Luke's DNR remains in place, but his parents still don't want to actively withdraw support. On rounds all stare at his X rays, then at his belly, now distended with gas. Roger suggests, "Why don't you put a larger tube in him?"

Amy, a cardiology nurse, says, "He was acting like he perfed again this morning."

"If he's perfed," Mee says, "it won't be good for him to be like that. . . . Put a larger tube in, but make sure you measure it so we don't go poking around down there. Frank, why don't you do that?"

"I'll do that," says Frank.

"That way we can blame you," adds Mee.

"Sure, blame me for the whole course!"

The crowd chuckles.

Mee sighs and says, "Never lost a case like this till Frank came."

More easy laughter.

And then one day, we're on rounds like any other day's—we've seen Ashley as usual, though Brandon, now successfully extubated, has been moved to the floor—and across the hall I see it. At first I don't completely understand. I wonder what they've done with Luke. Has he gone to the O.R. or the cath lab? His room is empty of all machinery, and in its place is an unmade bed, sanitary and unused, nothing wrong.

It happened late the night before, when withdrawal often happens, with the unit nighttime quiet and, not by chance, Elumalai Appachi the attending on call. Appachi, a forty-year-old doctor from Kanchipuram, India, fifty miles from the southern city of Madras, is beloved among patients' families for his genuine sensitivity and gentle manner, and out of the six PICU attendings, this young couple, Luke's parents, bonded only with him. Yesterday they asked to speak with Appachi and told him they were ready to withdraw their son's life support.

Some deaths are urgent—a cardiac arrest, say, when a code is called—but withdrawals are controlled. When the physicians have decided that further efforts would be futile, and the parents are in agreement, having had days to consider their options—only then will support be withdrawn. "I don't like to do them at all," says Appachi, of withdrawals. It's not so much the actual death he regrets, it's the anguish of the parents. "When the child is dead clinically, has no future, it's the family we have to take care of," he says.

And so everything is geared toward ensuring, first, that the patient will feel no pain, and second, that the parents will be comforted and supported during the withdrawal. A year from now Appachi may not remember the day a particular child died, but the child's parents will carry it with them for the rest of their lives. Appachi will ask if they want a baptism or have any other religious needs. Assisted by a nurse, he will gently direct the mom to sit and hold her baby. He will close the curtain of the glass room for privacy and then turn off the monitor so no one will see the child die on the screen (though another monitor will remain on at the desk to record the exact time of death). He will then very rapidly turn off all the drips, the medications that have been maintaining vascular tone and heart contractility, and slip the breathing tube out of the baby's mouth. Death tends to come quickly, which is why Appachi likes to have the live baby in the mom's arms before he turns off the medications and the ventilator. He leaves in lines through which he will administer fentanyl and Versed, drugs that will ease any distress the child may experience. Breathing will slow; the heart, deprived of oxygen, will soon begin to fibrillate and thus stop pumping blood to the brain and to itself; the brain will cease to function, thereby shutting down the lungs. Appachi will listen with a stethoscope, and when the heart has stopped, he will say, "It's time for him to be with God," if the parents believe in God— Appachi himself, as a Hindu, does, and so he often says this—or "She's at peace," or perhaps he will simply nod.

"We cry like crazy," he tells me. "I'm not ashamed to cry. We get attached to these patients. The day I stop crying, I go back to India."

He will then leave the room so the parents can be alone with their child. After an appropriate interval he will come back in and ask the parents to accompany him to a private room to talk while nurses clean the baby up. A "memory box" is opened and footprints and snips of hair are preserved for the parents. Appachi will make sure that the parents are stable and have somewhere to go, now, in the middle of the night. And if the cause of morbidity is uncertain, he will ask their permission to perform an autopsy. This can actually help the family grieve, learning the specific causes of their child's morbidity, and it can help the child's doctors understand what they couldn't while he or she was alive. When the parents have been taken care of, Appachi must call a hospital administrator, who will come to the unit and fill out the death certificate. Then he will contact LifeBanc, Ohio's organ-procurement agency, and the coroner. Later a transport team will arrive to take the baby to the morgue.

All of this happened last night in Luke's room, but it was the worst withdrawal Appachi has ever been through. It's one he will never forget, and one that will influence what he tells all future parents who find themselves in the same awful situation.

With the baby in the mom's lap and the dad beside her, Appachi turned off the drips and removed the ventilator tube, and Luke, who had never breathed on his own in all his six weeks of life, took a breath, and then another. He began to breathe normally. And kept breathing. Appachi watched, silently horrified. Had he been 100 percent certain about what had caused the baby's morbidity, he might have felt more sure of himself. But he simply didn't know—no one did. My God, he thought, have I made a mistake? Luke was breathing fine. Appachi began to sweat. The parents wept. Appachi knew the child couldn't breathe on his own for long, but he didn't even seem to be struggling for breath. Room air couldn't possibly have enough oxygen, however, to support him. And so as the baby boy continued to breathe, Appachi deliberately did something that likely prolonged the agony of his death: he held an oxygen mask near Luke's mouth, worried that Luke might be feeling the effects of suffocation. Twenty minutes passed. Twenty-five. A half hour. And at last, after nearly forty minutes off the ventilator, the breathing slowed. Appachi put his stethoscope to the child's chest, looked to the parents, and nodded.

. . .

"This is Julia Stone, day of life twenty-one," says Nihar Bhakta, a second-year resident who's midway through a month's rotation in this unit. He begins each presentation to the scrum in exactly the same way, having learned this rote format in med school. In this PICU he recites each patient's name, day of life number, and heart lesion, along with any significant historical facts, such as this baby's premature birth (she weighs less than four pounds). She was born with numerous lesions including a defect between all chambers of her heart—atrioventricular canal defect, in which the heart's central walls are missing—as well as other anomalies, the most serious among which is pulmonary atresia, the absence of a pulmonary valve or opening into the pulmonary artery.

Next Nihar will note any overnight problems before running down the front and back of a green sheet of numbers, organized by the body's systems. "CNS-wise," he'll begin, starting with the central nervous system, the patient's neurological status. It would not be unusual, for instance, for Julia Stone, as a preemie, to have problems with a hemorrhage between the two sides of her brain. Is her fontanel soft and flat (as it should be)? Is she on any narcotics for pain? (For an older patient, Nihar might specify whether he or she is awake and alert or still sedated.)

Then the respiratory system: most of the kids in here are intubated, so the resident reads off the rate of ventilation and all the vent settings, which break down into two general categories, pressure and volume. More questions to be answered: How much of each is the machine supplying? Is the patient breathing above the ventilator, or is the ventilator doing all the work? The ventilator will help Ashley Hohman, for example, if she tries to breathe on her own, but if she does nothing, the machine will give her twenty-two breaths per minute. Nihar ticks off the oxygen saturations, arterial blood gases, and pH counts, noting the acidity of the blood and the base excess, or the pressure of carbon dioxide in the blood, a description of how successfully the patient is being ventilated. After some more numbers—sodium, potassium, chloride, bicarb—he concludes this segment of his presentation with the physical exam:

how good air entry is, whether wheezes or crackles can be heard in the lungs.

Next comes the cardiovascular system: heart rate, mean arterial pressure, pressures in the heart's chambers into which lines have been inserted during surgery, central venous pressure. Taken together, these numbers draw a picture of the volumes of fluid in various parts of the cardiovascular circuit—whether in the pulmonary circuit (to and from the lungs) or the systemic circuit (to and from the body)—and how well the heart is pumping that fluid. Is the pressure in the left atrium low? That could indicate either bleeding or poor cardiac function. Is it too high? The left ventricle may be overfilled, or stiff, and unable to contract properly. The pressure in the right ventricle should always be lower than that in the left because the right needs to pump only to the lungs. If it's higher, it may mean that the lungs are backing up with blood. Many of these pressures can be chemically manipulated and fine-tuned so the body and lungs can be fed while the heart is allowed to heal and adjust after surgery. In each patient the system and its needs are unique.

"In congenital heart disease," Adel Younoszai explains, "you take this one system"—right side pumping to the lungs, left side pumping to the body—"and scramble the eggs, and you come up with a totally different system. You have to adjust your paradigm of thinking to that new system."

Because each patient with a complex heart presents a unique paradigm, much time is spent on this particular system during rounds. Nihar will usually pause here for a moment before moving on to the renal system, an indication to the surgeons that it's time to begin their daily debate. Here again, he finishes up with a summary of the physical exam: are the extremities warm and well perfused, what does the heart sound like, is there an appropriate "murmur" (caused by the shunt), or is there a "thrill" or a troubling "rub"? The subtleties of the physical exam are so great that it takes years of practice to develop competence, and some doctors will always be better at it, more closely aware of nuances and clues, than others.

Nihar now begins the assessment of the delicate, complicated kidneys. The amount of liquid going into the patient and the amount coming out provide a reliable reading of the patient's overall health and the perfusion of all the other organs in the body. Because the kidneys are extremely sensitive to blood flow and can be damaged by the heart-lung machine, peritoneal catheters are sometimes put in during surgery to dialyze the patient until kidney function returns. In such cases the nurses must monitor the patient's urine every step of the way; postop, some of the most important numbers a resident will read are the blood urea nitrogen and creatinine levels, with upward trends spelling trouble. The heart surgeons like a negative fluid balance, meaning more going out than coming in; healing tends to be faster and better when the patient is dry and there is less vascular stress. If there is a common cry among the speaking patients here, it's "Water"—uttered in the voice of the man crawling through the desert in rags—to which the response is typically a doctor's asking a nurse to swab the poor child's mouth with a moistened Q-tip.

Next on the agenda is FENGI—fluids, electrolytes, nutrition, and gastrointestinal status. By this point the presentation has been going on for a while, so a resident who focuses on what's important is likely to be well received. Nihar announces how much fluid maintenance each patient is receiving and what kind of nutrition is being fed intravenously. Nutrition is critical for the youngest patients on ventilators, who have scant reserves to draw on. And finally, the GI exam: can you hear the bowels working, is the belly soft or rigid, is the liver enlarged, blood backing up in the inferior vena cava, signaling right heart failure?

Then come hematologic issues, the blood's status and effectiveness in the body: white cell counts, polys, and lymphocytes, which indicate infection or the system's ability to fight it; whether it is anticoagulated; the H and H count, or hemoglobin and hematocrit, a measure of the blood's capacity to carry oxygen to the body. The lungs can be well ventilated, but if there's no hemoglobin to actually transport the oxygen, the body's tissues won't get fed. And last, a discussion of which tubes and lines can be removed from the patient's body or which ones need to be inserted.

These are the basics for every patient. When the resident has finished the presentation, he or she runs down the various medications being administered: diuretics, inotropes, pressers, or drugs to increase blood pressure; meds that do the opposite, dilate the vessels, such as phenoxybenzamine, nipride, and nitroglycerin; ACE inhibitors to relax arteries and promote renal extraction of salt; narcotics such as morphine, followed by methadone to get the patient off the morphine; beta-blockers to slow the heart and reduce pressure; anticlotting drugs such as heparin and Coumadin.

Mee, having felt the liver and examined the surgical site to see how well it's healing, will often fuss with the tubes coming out of the patient, leading over the side of the bed or down and over the foot of it, to make sure everything's draining properly. He's a nut about those tubes and all the physical gear that can kink, tangle, or otherwise not work right—because someone isn't paying attention or isn't thinking—and so screw the kid up.

Rounds begin in the X-ray room, where the group slowly accumulates, awaiting the arrival of Mee and his younger partner, Jonathan Drummond-Webb. The strapping South African has a light-brown complexion, short, thick brown hair, and a large mouth that he's fond of using. He talked his way into a fellowship here five years ago (when there was no fellowship program) and has since moved up the ladder, first to associate staff and now to staff and partner. Mee enjoys Jonathan's company, and he's grateful to him for working weekends, thus reducing the hours he himself has to spend at the hospital.

When the surgeons have arrived and are standing side by side in the middle of the cramped, hot room, the on-call fellow, often Makoto Ando or Frank Moga, begins by pressing the lighted buttons on the alternator to bring up the patients' X rays. Large lighted panels grind up and clank to a halt, and the group studies a series of pictures, discussing as they move down the line how much fluid remains in this one's lungs, whether the placements of that one's chest tubes and feeding tubes are correct, whether the diaphragm is in the proper position, the liver enlarged, the bowels filled with gas, whether that spot is a pneumothorax—spontaneous or from the chest-tube placement?—and any number of other visible indicators of improving or deteriorating health. Then the scrum squeezes out past the on-call room (bed, office, shower for the ICU attending) and into the first patient's room, which, this fall, is always room 7, Ashley Hohman.

Ashley was born on September 20 to Tim and Kelly Hohman, a couple in their early thirties, married for a decade. After trying un-successfully for several years to conceive, they began in vitro fertil-ization in 1998. Three attempts and three miscarriages later, they had initiated the process of adopting a child when Kelly discovered to her amazement and intense joy that their fourth and final in vitro procedure had resulted in a successful pregnancy: "Happiest day of my life," Kelly recalls. In the middle of her second trimester, though, they learned that Ashley had a heart problem that included a ventricular septal defect (VSD). At thirty-five weeks, her cardiol-ogist picked up signs of heart failure, and an emergency C-section was performed at Akron's Children's Hospital. Kelly heard a strong scream issue from her daughter and knew at that instant that Ashley was a fighter. Her heart was found to be gravely malformed, however, so the newborn was taken by ambulance to the Cleveland Clinic (a storm prevented a helicopter transport), where, once she had been stabilized for several days, Jonathan placed a band around her pulmonary artery, ligated her duct, and put in a pacemaker—none of which measures would cure her or even make her better, but only, it was hoped, keep her alive until a suitable donor heart could be found for a transplant. On the daily census report, the heart of Hohman, Ashley, is described as "CCHB, VSD, rudimen-tary ventricle, spongiform, noncompactation cardiomyopathy"—a rare conglomeration of defects that is best referred to as "complex."

Ashley, a baby on a ventilator waiting for a heart, has become her mother's full-time job. A slight woman with wavy brown hair, Kelly is staying at the Ronald McDonald House and arrives in Ash-ley's room shortly after rounds; she will remain here all day and well into the night, until she can no longer stay awake. Her hus-band, Tim, makes the hour-and-a-half drive north after work most nights, but every now and then he must return to their home to take care of the laundry and the bills. It seems almost surprising to him that you still need to do things like these when your longed-for first child is on mechanical life support. Both Kelly and Tim are in sales; she's on leave from a small publishing company, and he works for a technology concern where he now feels understandably distracted, and thus guilty for his lack of productiveness. He enjoys the drive to and from the clinic, though; he says it offers him quiet time for prayer and reflection.

Although Ashley scarcely moves all day, Kelly is always there for her, asking questions of the nurses so she can understand every bit of the monitor, the ventilator, the medications that are pumped continuously into her daughter in minuscule increments. She passes the time by reading or doing word-search puzzles, most of the time alone with one of the nurses and her unconscious baby, and in the front of her mind every waking moment is the ceaseless prayer for a heart. This prayer is in one way a terrible thing in itself, a wish many parents feel conflicted about: in hoping for a heart for their child, they are actively hoping for the death of someone else's. How can you wish for that? And how can you not?

Kelly will fuss over Ashley, brush her nose against her baby daughter's cheek, kiss her, speak to her. Kelly calls her Peanut and says, "I love you. Mom wants to hold you."

Ashley is every day the first order of business on rounds. A few days ago Tim and Kelly got bad, frustrating news from their surgeon, Jonathan: Ashley had developed fluid around her lungs, so much so that he couldn't transplant her. "I'll kill her if I try to transplant now," he said. So he had to take her off the list, he told Tim and Kelly, which meant that if that rarest of hearts came up, a heart for a five-pound baby, it would go to another child.

The following morning I asked Jonathan how the meeting with the Hohmans had gone. He closed his eyes and said quietly, "*Rough.*" In desperation and anger Kelly had said to him, "You're making a mistake." And when they looked at the new films in the morning, they saw that Ashley had in fact lost a lot of fluid—that is, the fluid in her belly and around her lungs had drained—and was breathing easier. Today, after the resident's presentation at Ashley's bedside, Mee asks, "Do we relist her?"

Kathy Weise, an ICU doc, replies, "She'll die without a transplant."

Roger, feeling a little frustrated, says, "If we're not going to transplant, we might as well not bother"—meaning, bother with all this effort of keeping her alive. "It's risky," he says. "It's salvage. But to me that's what transplant is all about." He wants her relisted.

Maryanne Kichuk, the cardiologist in charge of the transplant program, seems to be the one most against listing Ashley, though she doesn't round every day. After Jonathan relists her, Maryanne argues that Ashley's problems aren't only heart-related but include other organ systems, as well as a belly bloated with fluid, called acites. "A transplant won't fix her," she says, "and it may hasten her end."

Rounds continue, and sometimes they can be nearly as dramatic as anything that happens in the O.R. The scrum moves on to the baby in the next room, who was brought in yesterday with a conglomeration of defects: a double-outlet right ventricle (meaning that both great vessels are rising, improperly, out of the right ventricle), a VSD, and a pinched aortic arch, called a coarctation. He has gone into heart block, whereby the electrical signals that keep his heart beating have become impaired, and the external pacing wires that have been attached to him aren't working. Everyone stares at the monitors as the child, in PICU parlance, "tries to die"—right there in the middle of morning rounds. The room is crowded, and there is an urgent sense of commotion, more talk than action, as the irregular rhythm threatens to spiral at any moment into ventricular fibrillation, a fatal writhing of the ventricles as they cease to pump blood. The head electrician, cardiologist Rick Sterba, is paged and soon appears—but there's nothing he can do, and he says so. Jonathan speaks loudly and urgently to Sterba, gets right in his face and says, "*I'm between a rock and a hard place. I've got to get this kid into the O.R.*"

Mee asks, "Can we try pacing through the esophagus?"

"It doesn't work," Sterba replies.

The head echocardiologist, Dan Murphy, says, "I told him that."

Mee, aggressively serious now, informs Sterba, "*This kid's in real trouble if we don't pace him.*"

"Real trouble" around here, I'm quickly learning, means "death or near-death resulting in severe brain damage."

"We tried it on a hundred kids at Duke," Sterba says. "You can't pace the ventricle through the esophagus."

The emergency abates when the heart rate reverts to a more acceptable rhythm. Nevertheless, plans are made to get him into the O.R. "semiurgently." Jonathan is visibly agitated, and he and Mee continue to confer about the case after rounding on the last patient. "I'm still trying to figure out what I'm going to do in there," Jonathan says anxiously. Mee begins to say something but then stops, as if unsure of himself. Jonathan says, "No, that's why we're talking." Perhaps Mee doesn't know himself, would himself wait till he could see the heart and only then decide, as he often does; perhaps he thinks it's not his business to tell another surgeon how to handle a case; or perhaps he's trying to say to Jonathan, "In a matter of weeks you're going to be alone in Arkansas making these decisions—start now."

Jonathan has accepted a job as chief of pediatric surgery at Arkansas Children's Hospital in Little Rock; he's due to begin in mid-January. With Mee's blessing, he'll head south "to re-create what [Roger] taught me," he says, adding, "He's provided me with the tools to fucking rock."

By the end of the week Nihar will be starting his presentation, "This is Henry Parker, postop day three, status post–pacemaker insertion and coarct repair," and I'll have realized that both surgeons illuminated a facet of themselves during that scare over this patient. Jonathan said, "I'm between a rock and a hard place"—focusing on himself. Mee said, "This kid's in real trouble if we don't pace him"—his focus on the patient.

Jonathan, as I would soon come to understand, is a complex individual, almost impossible to read. He is loud, not reluctant to spar verbally when he doesn't agree, and willing to put people down in an if-you-can't-take-the-heat kind of way to advance his cause. Mee told me that Jonathan was a friend. He relieved him of a lot of work and weekend call and was a good surgeon. Jonathan can be ribald, quick to laugh, quick to tell a joke, intensely competitive.

He seems to me to have altogether too much energy. And I've never known anyone to be so doggedly conscious of my presence as a writer: he continually clues me in to the fact that people are posturing because I'm in the room, an accusation that I first suspected, and now am sure, as it's persisted, says more about Jonathan than it does about anyone else.

He was born in Johannesburg in 1959 and decided to become a heart surgeon, he says, the night he sneaked a transistor radio under the bedcovers and listened to news reports of the first-ever heart transplant, performed in Cape Town in 1968 by the South African surgeon Christiaan Barnard. The only child of older parents, he was sent to boarding school ("I had so much energy, my parents didn't know what to do with me," he remembers) and then attended the University of the Witwatersrand Medical School. He served in the South African army during the years that country was engaged in war with Angola. A practicing surgeon in his native city by the age of thirty-three, he moved to the United States to take a fellowship in cardiac surgery at the Latter-Day Saints Hospital in Salt Lake City, Utah, before coming to Cleveland to work with Mee in 1995.

This work, Jonathan confides to me one day in his office, glancing at pictures of children he's operated on, "humbles you, it makes you feel so small." The notion of this surgeon's ever feeling humble would raise many eyebrows, not least those of Mee's secretary, Debbie Gilchrist, who typically left Drummond-Webb's office mumbling to herself expletives that would have made her boss howl (and Gilchrist, Mee says, is one of the most valued members of the department for how much of her life she has put into this work and into this center).

Jonathan frequently speaks of Roger as his mentor or father figure; he says he loves this work and loves Roger, too. "There are some people you would help, there are others you would die for," he tells me. "I would die for him." They first met in South Africa, according to Jonathan, when Roger was speaking at a heart conference there, and Jonathan was immediately impressed by, in his words, "this little bald man smoking cigarettes." Since then his admiration for Mee has only grown. He says of him, "He's like a

Newton or an Einstein. He's a freak of nature. He's the Michael Jordan of cardiac surgery. He's the whole package." (Jonathan was one of at least three people I heard use the Michael Jordan analogy for Mee; several others likened him to Tiger Woods. Always, it seemed, it was a sports figure; surgeons are the jocks of medicine: they *perform*.)

Clearly Jonathan is a complex creature, but of one thing there's no doubt: he *loves* to be in the O.R. He *loves* to *operate*.

"I get such a *buzz* out of this," he tells me one day in the middle of rounds. "A day without surgery is like a day without sunshine! The day I wake up and don't look forward to coming to work, I buy a surfboard and a sack of marijuana and I go sit on the beach." In the O.R., with an open chest in front of him, as he's just about to perform the most challenging part of an extremely risky operation, he'll say, "The door is open, boys—let's go on in." In the operating room, as he says, he wants to fucking rock.

This kind of personality, much more than Mee's casual manner, is what I expected to encounter in a surgeon. It's often said that surgeons are arrogant, but it also seems to go without saying that if anybody has a right to be that way, it's these guys. The work seems founded on a premise of arrogance. To cut somebody open and root around in his or her intestines, repair a bone, or remove a cancer that's twisting around the spinal column; to insert a needle through a man's upper gum, behind his nose, and into his skull to suck stuff out; to try to fix the horribly cleft face of a child—all of these would be impossible without at least some degree of arrogance, but more likely a truckload. Being a surgeon carries with it an awesome responsibility. Arrogance, competitiveness, and brutal decisiveness may seem unnecessary or even abusive outside the O.R., but if they're what a surgeon needs in his or her psychological makeup to get the job done in the O.R., aren't they worth the price? However much physicians and nurses may gripe about "difficult" surgeons, few seem actually to begrudge them these traits, given the nature of their responsibilities. A powerful ego may be especially necessary for those who choose to operate on the hearts of children, and who thereby bear the weight of entire families that hang their very being on the surgeon's personal ability to fix their babies. This happens almost without exception, in every case. For Jonathan that means 150 to 200 families a year, and for Mee between 300 and 400—hundreds of families a year in here, family after family after family. Anyone on the staff will attest to the fact that though peds heart surgery is a team effort, the parents' bottomless need is nearly always focused on the surgeon.

At the top of Jonathan's list these days are the Hohmans, who want nothing more extraordinary, nothing more ordinary, than to bring their baby girl, Ashley, home. Theirs is a comfortable house, only two years old, with a thick lawn and an attached garage, on a cul-de-sac in central Ohio, not far from where Tim and Kelly both grew up. They met when Tim was in his first year at Ohio State; he was best friends with Kelly's sister's boyfriend and somehow got saddled with the errand of picking up kid-sister Kelly, then just sixteen, from a high school football game. They ran into each other again at a party, began to date, broke up after two years to explore other options, reunited, and eventually got married. After three years of married life, they began to try to have a baby—the start, Kelly says, of their "long, urgent need to be parents."

Overtly proactive regarding Ashley's condition, she is eager to talk about personal matters with anyone who enters the room. Speaking with me is a way for her of affirming their choices, of sending a message out into the room that they intend to do everything they can to help their precious daughter live, despite the enormous emotional, spiritual, and financial toll, despite the uncertain outcome. Not only do they not know whether Ashley will survive, but they can't know what kind of life she will have if she does: has her brain suffered, will she have terrible lung disease, will she need long-term dialysis? They speak candidly about their lives, as if to underscore the confidence with which they have made the choice to try to save their baby girl.

That choice began with four in vitro fertilization attempts, an increasingly common story today and a big investment, both from a financial standpoint and from a physical and spiritual one. Even at the most basic level it involved "a lot of pain," Kelly notes: she had to give herself regular shots in the belly and thigh.

"I had to think: How important is this to me?" she says, recalling her six years of trying to become pregnant. "In my heart I wanted so badly to give a child—whether it was our own child or an adopted child—our love. We had so much love to give. We wanted a child so badly. And to bring out the kid in us, to rejuvenate Mom and Dad."

When the fourth in vitro took, the couple was thrilled. Assuming, of course, that it would be a pregnancy like any other, they made the usual plans: Tim painted the nursery, and Kelly decided that she would return to work after taking a brief maternity leave. But their life now pivots on the day a doctor performed an emergency C-section on Kelly to save the thirty-five-week-old fetus whose heart had begun to fail in her womb.

"Going through the things we have gone through," Kelly says, tilting gently back and forth in the wooden rocker that is her main home for fourteen hours a day, going on sixty days now, "we're able to appreciate children a lot more. And appreciate where they come from and how precious a gift they are. And how precious life is."

Kelly is mostly by herself all day. Often Ashley squirms and opens her mouth and appears to cough, but no sound comes out; with a breathing tube inserted just past her vocal cords, she is absolutely silent when saliva and fluids begin to clog the airway, the constant obstruction requiring suctioning many times a day. Most evenings Tim drives up to join his wife and hover with her at their daughter's bedside. With his short hair going to gray, glasses, and mild manner, Tim seems very much a midwestern Everyman, except that the mild manner is just a little *too* mild. He sometimes has the overly placid look of someone who's about to snap—or who has *already* snapped. The monotone of his voice gives me the sense that he's on the edge of becoming unreasonable. Skepticism and doubt creep into his conversation: he appears to believe that the doctors have their own agenda, an agenda that has nothing to do with his or his wife's wishes. Marc Harrison notes that he and Kelly have exchanged unusually sharp words. Tim makes statements like, "I get up in the morning, so that means I got to bed the night before." It's impossible to know what his state of mind is, but who could blame him if he *were* losing his grip? He and Kelly talk about Ashley's distant future. Is this a naive fantasy, or a way of feeling they're taking an active role in her recovery, a hope that positive thinking may literally help Ashley's body? But it's one thing to be positive and hopeful, and another thing entirely to speculate about the Christmas presents she'll ask for one day (she'll definitely want the game Operation, Kelly suggests). Perhaps it's simply more than this couple can bear to think that Ashley Grace Hohman won't come home one day, that her life will end in this room, and that they'll have to drive home alone to their clean, warm house on the cul-de-sac with its painted nursery and its dense lawn. Edgy-calm though he is, Tim is nevertheless articulate and reflective. "Sometimes I think, Why me?" he says, leaning forward in his chair next to Ashley's bed. He pauses. "And the only thing I can come up with is, Why *not?*"

Kelly stands and hovers over their baby and speaks to her softly. Ashley's mouth is hidden by the tube and tape, but her eyes clearly lock onto her mom's and stay there.

Ashley was relisted for a transplant two full days after she was pronounced ineligible. During that time a heart just right for her may have become available, but because her name was taken off the list maintained by UNOS, the United Network for Organ Sharing, the Hohmans will never know. With Ashley back on the list, though, Kelly can speak warmly of her surgeon once more and is quick to repeat Jonathan's comforting words to her: "Ashley is the last thing I think about before I go to sleep at night," he said, and she keeps that always in mind. Kelly lives her days praying for a heart and praying that Ashley won't suddenly get sick or become septic—anything that could take her off the list again. Her spot on that list represents Ashley's only chance of staying alive, and when Kelly drifts off to sleep each night, her main hope is that she'll be awakened by a phone call telling her a heart has been found. But that's all she can do, really: hope. She doesn't need to be here for any practical reason other than to look into her daughter's eyes. Day follows day, week follows week.

Standing at Ashley's bedside when Kelly is not in the room, Jonathan is less comforting. "She's not a rose," he says, eyebrows raised. And another time, "Ashley. Our little bundle of joy." As he awaits Mee's arrival in the X-ray room for the start of rounds, Ashley's pictures up on the lighted board showing her belly bloated with ascites, he shakes his head and says with a sigh, "She needs a heart."

Steve Davis, Diet Coke in hand, replies, "She needs a trip to heaven."

And softly, Jonathan says, "She looks like she's going that way."

Once Mee is present and all films have been reviewed, the scrum gathers around Ashley's warmer bed. The resident runs through each system, presents the gases, the pressures, the drips, the vent settings. The discussion centers on Ashley's belly, now swollen with fluid. Mee believes it's due to her rotten heart, to the fact that it just doesn't pump very well. The other issue is her head, which hasn't grown much since she was born; Jonathan relisted her only after an MRI confirmed that she wasn't bleeding up there. He says he'll take some of the fluid out of her later this morning. Returning after rounds and donning gloves and mask, he puts his hands on the baby, explaining, "I need to turn her a little on her side so the fluid is toward me," then says to Ashley, "Oh, you're not going to be happy with me."

The PICU attending, Elumalai Appachi, steps into the room to join Jonathan and the nurse (parents are asked to leave the room during procedures). He asks, "Do you need me?"

"Always," Jonathan replies. "OK, nunu," he says, inserting a needle into a belly that's stretched shiny and tight like a balloon that could pop at any moment. "It's all right, I'm sorry," he says. Ashley would no doubt scream if she could, but her gaping mouth is silent, no breath passing over the vocal cords. Jonathan hooks the needle up to a tube fitted with a plunger and a two-way valve. Dawn Blair, one of Ashley's regular nurses, pulls the plunger, drawing out fluid. Jonathan asks Appachi, "How much do you want to take off?"

"Take off a hundred cc's," Appachi says. "There's maybe five hundred or six hundred in there." When the big syringe is full, Dawn says, "There's twenty cc's." She flips the lever on the valve and expels the thick yellow fluid into a holding bag, then flips the lever back and starts drawing again.

Ashley appears to struggle a bit, and Jonathan says, "She's getting a little personality."

Dawn smiles and says, "Yeah, she starts banging at your hand when you're suctioning."

Ultimately Dawn removes 260 cc's of fluid from Ashley's belly. After Jonathan thanks her, Appachi asks her to send some of the fluid to the lab for analysis. Then Dawn, who clearly has become emotionally attached to Ashley over the past two months, turns to Jonathan and says, "I wanted to ask you, what are the chances that you'll find a heart?"

Jonathan doesn't look at her. He answers only, "I don't have a crystal ball."

Later I ask Maryanne, the head of the transplant section, about Ashley's prospects. She tells me that one in four infants on the list dies before a heart can be found.

The next day, as the rounding doctors leave Ashley's room, discouraged that her belly has filled right back up, Dawn looks to Frank Moga, the surgical fellow from Minnesota, and asks, "You can't fix her?"

Frank says, "Not today. But I'm smelling a heart. Soon. I'm sensing a heart." He glances at me and flashes a characteristic grin, teeth clamped, then says, "*Voodoo.*"

The following Sunday Frank is paged at three in the morning. A page can mean that something's up at the hospital with an inpatient or a newborn in need of emergency surgery. Before prostaglandins came into wide use in the late 1970s, this happened all the time at congenital heart centers—kids with a variety of defects needing to be cracked open at once, whole cardiac teams roused from sleep to get into the hospital fast. Because prostaglandins help maintain a fetal state, such urgency has become the rare exception, most often seen now with a defect called total anomalous pulmonary venous return (TAPVR), in which the pulmonary veins—four vessels that return blood from the lungs to the heart's left atrium—are blocked. At birth, blood and oxygen flood the newborn's lungs, but with TAPVR the blood has nowhere to go, so the lungs fill up and the baby immediately gets, as one cardiologist puts it, "sick as stink." But Frank's page, it turns out, isn't for a TAPVR surgery; it's for a heart harvest. Frank likes to get hearts, especially for patients who need them as badly as Ashley does; the wait has had the whole team wondering when an organ will come. But when he asks who the recipient is, the transplant coordinator says a name he doesn't recognize. Frank suggests there must be some mistake, but no: it's a young adult with a congenital heart defect who's also on the list. The news is deflating.

"We're getting hearts for everybody but the kid who needs one most," Mee remarks to Frank during rounds a few days later. To me, Frank says, "I *did* smell a heart—it was just the wrong one." Implicit in this, of course, is the suggestion that he does have voodoo powers, and they're working.

The subject of voodoo comes up often here, maybe because so many of the decisions made about the center's cases seem to be intuitive and so many events can't be explained. A kid will be sprinting toward the end zone, as they say, and then for no apparent reason he or she will turn around, stay alive, and prove everyone wrong. That's what happened, for instance, with a little preemie named Julia Stone, a four-pound baby born with terrible problems: AV canal, pulmonary stenosis, aortic atresia. Like so many others here, she is the desperately hoped-for child of wonderful parents whom the PICU staff adores. Mee addressed her condition by putting in a Blalock-Taussig (BT) shunt. One of the original palliatives for heart disease, originally developed in 1944 to help patients with tetralogy of Fallot, a little tube not much bigger than a cocktail straw is first sewn into one of the arteries rising off the aorta and then attached to a pulmonary artery. In patients whose pulmonary artery or valve is blocked or nonexistent, the device channels blood to the lungs.

In this case, the BT shunt was only a stopgap; the hope was that Julia would grow enough to enable Mee to attempt a more complete repair. (Any baby under two and a half kilos is considered high risk, and this one is under two kilograms.) But after the shunt was put in, aortic stenosis prevented enough blood from traveling up the aorta, so the baby was brought to the catheterization lab, where doctors would try to open her pulmonary valve with an inflatable balloon. The first balloon didn't do enough, was too small, and the second proved to be too big: they blew the baby's valve. Her white-cell count went up, and her cardiologist feared sepsis;

for those reasons Mee could not be called in that night to operate. During rounds a few days later Mee confides to a surgeon who's come to interview for Jonathan's spot, "I thought we were going to lose her on Friday, but she hung in there." Appachi, the attending on rounds today, tells the group, "The parents are still hopeful—unfortunately." There's nothing left to do now but wait. Mee, too, remains hopeful.

By the end of the week the baby has stabilized. Roger announces during rounds, "If she becomes a candidate, we'll do her."

Kathy Weise asks, "What does that mean?"

It means getting the fluids down and avoiding sepsis, Mee says. On hearing this, both Jonathan and Frank turn away from the group, shaking their heads; they can't believe he's still talking about operating. Jonathan whispers, "You've got to let her go." Frank looks at me and says, "This is what you shouldn't see. This is the horror. She's dying a slow, horrible death."

I say, "Roger apparently thinks there's hope."

"If monkeys fly out of my ass," says Frank.

Later I ask Mee himself about it, and he makes two points. First, the parents have to be the ones to decide whether to pursue aggressive treatment or withdraw life support; otherwise they'll spend the rest of their years wondering if they were wrong to let the doctors make their decision for them. And second, as to when it's best to withdraw support, Mee insists, "There's no easy formula for this, and there never should be."

To most everyone's surprise, the baby doesn't die; she's still there when the scrum stops in on Monday rounds. There's a chance that Mee may be able to operate, and if he can operate, he thinks he can fix her heart enough to allow her to recover. On Thursday he asks Appachi, "Have there been any setbacks in the last twenty-four hours?" Appachi spreads his hands apart like an umpire indicating "Safe": No, nothing.

The next day, Friday, Mee looks up from where he's standing beside her warmer bed and says, "Well. I'm thinking Monday."

I catch Frank's eye and say: "Monkeys." He grins and says, "Right! She's gonna live. Roger's voodoo is stronger than my voodoo!" Frank's grin is not just a smile; he looks as though he just stuck a finger into a light socket—and was *liking* it.

He happens to be well rested now, and he has call tonight, which is why he'll be here tomorrow morning, Saturday, preparing to put a new line into this baby, whose blood pressures have just dropped precipitously. Sepsis, or toxic blood, can kill a baby, and Frank tells me now, "She's trying to die on us."

He hopes that the new line will help, not only by providing fresh access for the delivery of meds but by replacing the old line, which may be what's causing the infection. As he's finishing up, having gotten a line into the right subclavian—a small vessel, no mean feat on an infant this tiny—Jonathan appears and asks if Frank wants them to start rounding on the floor, rather than in the PICU. As on-call fellow, he's responsible for beginning rounds in the X-ray room, reviewing films and overnight events.

"No, I'm almost done here," he answers.

When the sparse Saturday crew rounds in Julia Stone's room, Jonathan says, "Someone should get the parents in here, let the mother hold her warm baby."

No one thinks the girl will live. Kathy Weise asks Jonathan what she should tell the parents. "They'll ask if there's any chance of operating," she says. "And you know what Roger will say." These latter words are spoken hopelessly: because Julia is septic, everyone knows that no operation will be possible. Again I look at Frank, and he says, "I'm not making any predictions," though he's not hopeful. "She has a will to live. I wasn't backing her before; I stepped back. She proved me wrong. Now I'm backing her."

During Monday rounds the scrum passes by Julia Stone's room, which is now empty and clean. It's a deflating, depressing sight, this empty room that once seemed so palpably filled with hope and prayers and intense medical management and energy and thought and debate—so much life. Where has it all gone? It's as if someone pulled the plug and it all drained out.

At the PICU desk Marc, Steve, and Frank discuss the withdrawal of support. Frank says the baby girl would have had a difficult life, with a bad heart and a lifelong need for open-heart surgeries to replace valves.

Marc recalls caring for another fatally ill infant, whose parents were Wyoming ranchers. "The father said, 'When a calf is born this way, the mother licks it dry and walks away. And I think that's what should be done with this child.' " He pauses. "I think people who work close to the earth seem to sense better that some things aren't meant to live."

Steve Davis winces, says to Marc, "You're getting too philosophical," and wanders off.

Three doctors, all roughly the same age; three different defense mechanisms. This is life in a pediatric intensive care unit, and it is normal. Of the five hundred cardiac cases they'll see here in a year, ten will die, on average. The majority of them will die slow, horrible deaths, and the rest will die quick, horrible deaths, and all the parents will suffer, and all the doctors feel it. The whole unit takes a blow when a child dies. You probably wouldn't notice it unless you worked here, and maybe not even then, but I can sense immediately when something grim has happened. There's no absence of talking, nor any lowering of the routine noise of the unit, but there's an absence of unnecessary talk, a certain quiet where there would ordinarily be everyday banter or chat. It feels as if fewer people were actually working; I think it's because everyone retreats into himself or herself a little, so personalities take up less space. If there's anything unusual about this day, it's the concentration of so many seemingly hopeless cases on the floor. And another is on its way. By the end of the day, a once-clean and sterile-feeling room will be occupied by a new family, a new infant named Drew actively dying of yet another variation on the theme.

It feels endless.

4. Scaramouche

"You can't hide when you're a peds heart surgeon. You know who you are and where you stand. Because the stakes are so high, because there's so much at risk, there's no room for dishonesty. I've got to be so sure of myself—you know, know my essence. It's clear. It's beautiful. Elegant. Brutal. There are not many jobs where you're forced to know exactly who you are. You can't lie to yourself here, because if you lie to yourself, it becomes very obvious. Somebody dies."

He says it as though it were hilarious but undeniable: "It's fucked in the head—that's the truth of the matter." Frank Moga has given it a lot of thought.

"They don't teach peds surgery for a lot of reasons, but part of it is that it's the cardiologists who teach, not the surgeons. You've got a monthlong class in cardiology, just four days of which will be peds, because even though peds is more complicated than adult, heart disease is the number one killer of adults; that's where the money and the numbers are, so that's what the focus is. But for those four days when we did congenital, that's when it occurred to me: I *understood* peds. They show you these diagrams of, say, transposition, and you can calculate the ratio of the shunt flow, and you can calculate what's going on in the heart from the angiogram numbers, and I just saw how elegant it was, as far as plumbing. It can be confusing at first, but it wasn't for me, I just got it right away. And people in my class would say, 'You don't want to do this. Peds heart surgeons? They train like five years in general, then three more in cardiac, and then they have to keep training for peds. It's the most ridiculous thing in the world to be a pediatric heart surgeon.' So of course when people say it's too hard, that you'd have to be fucked in the head to do it, well, my personality is to say, 'I gotta give it a roll!' "

Francis Xavier Moga, the second-youngest of nine children, grew up in Lake Forest, Illinois, and attended medical school at the University of Illinois, graduating in 1991. (He's a former rugby player, like Roger Mee and Jonathan Drummond-Webb.) After first looking into a residency at the Mayo Clinic, in Minnesota—where, he recalls, all the docs wore suits, and the carpet was so clean you could eat off it—he drove his little convertible Karmann Ghia out to visit a community hospital in Minneapolis, the Hennepin County Medical Center. Unaffiliated with any academic institution, it promised him the opportunity to practice hands-on medicine. As he pushed through the entrance doors, a drunk began to vomit all over the lobby floor. Frank observed the scene. No one seemed to pay much attention; the hospital staff walked around or stepped over the mess. This kind of hospital setting was more to Frank's liking—his was a front-line temperament.

During my first three months at the Cleveland Clinic—my introduction to peds heart surgery, a time when there were so many sick kids in the unit, and each morning I learned that no heart had come for Ashley Hohman the night before—Frank Moga was my interlocutor, my interpreter and guide on foreign soil, the man with the machete who cut through the jungle brush to lead me to the orchid.

It happened naturally, because Frank was all but in the mirror when I looked at myself. He was about my age, was also from the Midwest, and had likewise been a liberal-arts major in college, in his case St. John's, one of the last males-only colleges in America (my own first book was about all-male education). When he saw me arrive for morning rounds, he'd say, "Take that stupid thing off, man," because the tie I'd had since high school looked just as silly and uncomfortable now as it had then—just like the tie *he'd* had since high school. Same with those Johnston & Murphy shoes, a pair of which he also owned. I think I spooked him a little: he knew exactly who I was, because he'd come from the same place, at the same time. But more important, he understood what I was doing, what I needed, in a way that most people don't when a writer starts hanging out with them. I walked into the O.R. one day while he and Fackelmann were opening. I was still an unknown quantity to Fackelmann, who said what most people say, or at least think, at first: "It's the writer." Frank was dissecting out a kid's heart. His words were muted by his mask, his louped glasses held on by a Rasta-colored band over his bouffant cap. "It's our conscience," he corrected Mike. This was true mainly because he said it was. He was almost writing parts of my book for me by thinking out loud. In itself this wasn't that unusual—it sometimes happens when the people I'm watching are especially articulate about their work—but Frank uped the ante: he *knew* he was doing it, knew what I was after, the private language and hidden ethos of the world of pediatric heart surgery. All med students and residents are more or less trained to be hyperaware, but he was hyperconscious of his own capacity for it. He'd thought about it, and he didn't want to contribute to one more phony medical-miracle, surgeon-with-the-hands-of-God story; he'd assessed the situation and decided to participate; so he more or less told me, If you want to see something interesting, if you want to see what this is *really* all about, I can show you.

Frank was literary the same way I was even though he was a doctor and a surgeon. His bookshelves at home were loaded with books I also owned, but his were better: in addition to the podium-size *Atlas of Surgical Procedure, Textbook of Surgery,* and *Principles of Medicine,* he had works by Raymond Carver, Borges, and Jim Harrison, as well as signed first editions by Bruce Chatwin and Mark Helprin. A doctor had to train long and hard, on crazy schedules like forty-eight hours on, forty-eight hours off, and a heart surgeon had to train even longer—fifteen years before getting to do the work he or she had set out to do. That Frank should have any kind of literary taste at all was somewhat odd, but for me lucky. His musical tastes were just as eccentric, running from Burt Bacharach to the Grateful Dead. He drove an old green pickup, which seemed to be the surgeon's ego speaking; I could make fun that this was just too cool, the heart surgeon who drives a pickup, aren't we the maverick?—but the real reason for it was that he couldn't stand being a cliché, like Jonathan, who had a shiny black BMW and wore natty threads.

Frank spoke a beautiful language thanks to his literary bent. For him a heart was not just a four-chambered pump; it had atria (*"nice distensible bags"*) on top of the ventricles (*"beautiful muscles"*), and he relished the thought of it, this incredible organ. He clearly loved its elegance, the art of it. I tried to ask Frank as many questions as I could without abusing his time. I could have asked just about any other doc what purpose prostaglandins served, for example—this natural chemical that has transformed the world of congenital heart disease—and received by way of an answer something like, "We give prostaglandins to keep the duct open," or an equally dry alternative. But when I asked Frank, "What are prostaglandins?" his reply had a sort of succinct comprehensiveness that told me all I needed to know: "Congenital heart disease is really well tolerated in utero. Prostaglandins keep the baby in that fetal state after it's begun to breathe oxygen."

Or again, one day when the scrum had left a patient's room and was heading to the step-down unit to finish rounds, I asked Frank about a word that'd been used to describe an infant's condition: "What's it mean to be acidotic?"—in this case it was lactic acidosis. "What's happening and why is everyone so worried about it?"

"When you run a marathon," he said, "your body produces acid—you know that, that's why your muscles ache. Well, that kid's body's been sprinting all night long, and it's still sprinting hard, and that's not good. He can't keep doing that. I'm gonna jump in here and grab a cup of coffee."

As a matter of routine at this point in rounds, he'd hit the blue overhead plate that opened the doors to the two operating rooms, take an immediate left into the small surgical lounge, and grab his seventh or eighth cup of the morning, then catch up with the scrum

again on the way down the hall to the "ward," as Mee called it, or
the step-down unit, where kids who no longer needed intensive
care finished recovering before they went home. There were
twenty-two beds here, and fewer than five hundred admissions a
year; the eight beds in an area known as the pod—a semicircle of
curtained-off spaces surrounding a glass-partitioned nursing station, with all beds in direct view of the nurses—were reserved for
kids who needed closer scrutiny. The rest of the beds were divided
among eight private or semiprivate rooms. The care was not as focused or intensive here, simply because all the patients weren't on
the verge of dying, the nursing seemed more erratic, and the residents who presented the cases on rounds were first-years, just out
of med school. It was their first time taking care of cardiac patients,
so they struggled to read numbers off sheets and were obviously
foggy in many areas. Mee, in his perpetual black slacks, coat, and
tie, was likely to do some teaching here. To one hapless resident, for
instance, he said, "Instead of just reading off the sheet, you should
write the story." A presentation should give some insight into what
was really going on with a patient, in other words, rather than being
a mere recitation of electrolytes and mean arterial pressures and
temperatures and medications administered. "You should write
notes," he advised. "That will help you synthesize salient facts, to
put it all together. I don't mean to be insulting, but my daughter
could present that. I want you to think, not just produce numbers."
The resident stood and took it—it had happened before and would
happen again, all part of the training—until Mee left. Frank made
his eyebrows jump, then shrugged; he'd been Roger's punching bag
in the O.R., so he knew how the first-year felt. Didn't matter; it was
kind of funny—he had no real stake in this world. He'd be leaving
soon, to return to his practice in Minnesota.

"I came here because I needed a pedigree," he said. "I came here
specifically because I needed to learn the switch, because Roger is
the best in the world. Roger's got a reputation for being difficult,
but that's part of the culture. It's there from the beginning. In med
school they break down your personality—it's socially acceptable
brainwashing. It changes when you sleep, when you eat, how you

think, who you are. You're glad when you get off with a hundred-and-twenty-hour workweek—that's a good week. They destroy you, then they build you back up. We used to watch *Full Metal Jacket* and howl, because it was the same thing! That is exactly what it's like to become a surgeon. You have to live it. It takes total commitment. You're a different animal."

"What do you mean, 'different'?" I asked.

"It's active; you fix things. Cardiologists talk about it, but surgeons fix it. Just the other night my girlfriend told me I bolted upright, middle of the night, said, 'Whaddaya mean you can't fix it?! Givittame!' and went back to sleep. You fix it. That's who you are."

Frank talked, and I loved to listen. I could listen to the guy all day.

"Frank," I asked, "when you're in there assisting, are you learning it? What are you *doing?*"

"The difference between assisting and actually doing the surgery, it's the difference between watching pornography and actually having sex. Same thing. The first time you're in there you're just like a virgin with your pants around your ankles. You may think, in your head, Oh, man, I'd love to be in bed with her, and you can see how it would all happen and how beautiful it would be. How you'd make love. Right? And then suddenly you're there with her, *naked,* it's about to happen, and you think to yourself—*Whooooaaa, this isn't how I imagined it!*"

And he *laughed.*

I walked with him every chance I got, just tagged along. If he was eating saltines in the O.R. lounge—there were boxes and boxes of cellophane-wrapped saltines and graham crackers in there; when Frank walked in once and saw me munching a stack of saltines, he said, "You're having my favorite lunch!"—I'd veer from my intended course and sit down and ask him a question.

On another occasion I was accompanying him down the corridor toward the PICU, and he stopped short of the doors. I'd been telling him about working with a chef for whom the notion of perfection was an impossibility that kept driving him. I asked Frank, "Do you search for the same kind of perfection in the O.R.?"

"A surgeon's quest is not for perfection but rather for grace," he said without hesitation.

"Define grace."

"A moment of clarity. You're not yourself, you're an instrument of something else. When I was in thoracic surgery, doing a pneumonectomy, taking out part of a lung, I had to sew the pulmonary artery back together with one hand because I was holding the trunk of it with my other hand. I didn't even think. Forehand, *boom, boom, boom, boom,* four backhand, and it was done. The guy I was training under said, 'Frank, that was a moment of grace. Never forget it.' I hadn't realized, I was surprised, but he was right, and to this day it delights me to think about it. A guy named Pat Ross said that. That's what I strive for.

"Larry Bird was once asked about Michael Jordan after a game between Boston and Chicago when Jordan was on—really on, even for Jordan—and Bird replied, 'That wasn't Michael Jordan, that was God.' That's when you know you're there—when it's not even you."

One morning on rounds we went up the stairs at the back of the PICU and wound our way through the maze of corridors and stairways to the adult unit on the fifth floor of the G building, where adult cardiac patients recovered. Jonathan had to check on a congenital heart patient he'd operated on, but Frank didn't need to go to the bedside, and so we waited outside in the hallway. I always had some question ready about surgeons or pediatric heart surgery; today I asked him which surgeons were considered the best.

He shook his head dismissively and said, "These guys cultivate their own mystique, their own following. It's an attitude. It can be almost pathological. They construct their own legend. Roger Mee walks on water here. You go to Michigan, they'll tell you, 'Ed Bove walks on water.' That's why Jonathan's leaving."

"Why?"

"He can't walk on water here."

"So someday *you're* going to walk on water, then?"

"No way, man. I'm a swimmer all the way."

And then it came, as it did frequently throughout the day: his laugh, a *hee hee hee*. His eyebrows would bounce and jiggle almost independently of each other. Then his lips would part to reveal clamped teeth, and then a throaty *Haaaa!* would come out like sandpaper, and he'd say, "Right, man, I know! Exactly!" Like he was reading my mind. And sometimes I thought he was.

The bug-eyes, the hair straight out of a 1970s sitcom, the laugh. *What's this all about?* The wigged-out mad grin, the *Isn't-this-amazing? Isn't-this-hilarious?* expression on his face, *I-know-man-can-you-believe-it?* But we weren't at a carnival midway, we weren't watching a freak show, and we weren't in Stanley Kubrick's War Room; we were in a PICU, where all the kids were sick as stink and some were about to die, and Frank Moga was grinning like a maniac. The guy was *alive,* directly plugged into the life source.

"My mom used to say I reminded her of the first line of *Scaramouche,*" he said. " 'He was born with a gift of laughter and a sense that the world was mad.' This is madness. *Hee hee hee.* Right, man. You look at that! It's madness."

It *was* awful. Jonathan was recannulating Jessica Bojarski right in her PICU room. He had closed the room off, shut the door; everyone was scrubbed, and now he was reopening her groin, spreading the muscle apart to dig out the femorals and redo the lines to stop the bleeding. All so this nine-year-old child of anguished parents could stay alive long enough to get a new heart. Jonathan had transplanted her just two years earlier, but the coronaries had already started to block off, one of the major complications of a transplant. At age seven, in August 1998, Jessica had become ill with what everyone assumed was a stomach virus after a camping trip; when she didn't get better, her family took her to a hospital, where doctors diagnosed congestive heart failure and arranged for an immediate transfer to the center. Here she was found to have dilated cardiomyopathy, a disease of the heart muscle itself. She'd likely had it always, but neither her mom, Tammy, nor her dad, nor her stepdad, nor any of the rest of the family had ever had the slightest inkling that anything was wrong. The family's whole life pivoted on that month of August two years ago. She'd been listed for transplant and gotten a heart, but according to Tammy, "There were problems with that heart from the get-go" —mainly a bad course after the operation and some early signs of rejection. Now Jessica's arteries were clogging off, she'd developed post-transplant lymphoproliferative disease, and she'd been waiting in the hospital for weeks for a second heart—waiting in a room appropriately decorated for a nine-year-old at Halloween. She was status 1 on the list, but no heart had come yet. Then one night—"It was God working," Tammy said—Appachi left the PICU and took a stroll through the ward. He'd decided to check in on Jessica—not for any particular reason; she wasn't his patient anymore—and he just happened to be on the ward, just *happened* to be standing at the front desk staring at the child's monitor when she started to Brady, her heart slowed, the rhythm slowed, and he realized she was arresting and ran to her room. Because Appachi was right there, because he had his eyes on the monitor *the second* things began to go wrong, he was able to resuscitate her quickly, with ten compressions. Jonathan got in there in the middle of the night and put her on ECMO, the extracorporeal membrane oxygenator, a twenty-four-hour heart-lung machine that's hell on the blood and the body. It's so brutal, in fact, that a kid can be kept on it for no more than a few days or so before being destroyed by it. In Jessica's case, Jonathan told Tammy, the limit would be two days. And so here she was, this tough, unconscious little girl who had gone into full cardiac arrest and then been crashed onto ECMO, and there still wasn't a heart for her. Meanwhile, her entire family was huddled in the lobby out by the elevators, all of them red in the face from crying and scared to death that this time Jessica was really going to die, that after two years of struggle, hospitals, fear, and pain—here it was, they were going to lose her.

Earlier this morning cardiologist Rick Sterba and the nurses and Jonathan had all stared at her monitor in dismay. Jonathan was wearing his surgical whites and had a hood and headlight on, intending to recannulate after rounds. Sterba all but whispered, "You *listed* her? I can't believe you listed her."

Jonathan said, "What do you want me to do? Leave her to die?"

Sterba paused briefly, then quietly said, "*Yeah*." Meaning, that would be the humane thing to do, end her suffering, not prolong an awful death by waiting till ECMO has destroyed her kidneys, her lungs, her brain.

Jonathan headed for the film room to begin rounds. "I'm going to find her a heart," he said. "I'm an optimist. If I can't find her a heart, I'll find some device to pace her. I've been with this family since I transplanted her in 'ninety-eight."

"And that other little kid," Frank went on, "Julia Stone, look at her! She is *melting* into the pillow! This is the horror, it's madness."

And then Frank's scared, horrified look transmogrified into a look of hilarity—or maybe they were the same thing—and he laughed his easy, hearty laugh—he'd been in this world for years—and laughed some more, and then, his face losing all expression, he took a breath and said, "As long as you don't lose your love of the family. There's an incredible amount of comedy in this world, and a horrendous amount of tragedy, and you don't want to let the tragedy define what you do."

Frank wanted to be connected to it all, plugged in to all the horror and the beauty, and still stay human.

Some people become inured to the horror—trauma surgeons, who see so many horrors they don't even notice it anymore; *nothing* fazes them. Some get that way out of necessity: their psyches won't allow them to invest themselves emotionally in all these situations and still function as doctors, do the actual physical work—it's simply too much to bear. Frank bore it in his own way, by seeking out the potential art and beauty and grace in this world—that was how he dealt with the suffering that inevitably surrounds all who choose this work. And it was why he loved to go on transplant runs, or harvests, as they're called. He loved it—anytime, day or night, just page him and he'd be there. That was the shit, man. "*It's so life-affirming,*" he said. Adel Younoszai called him Farmer Frank because he was always the fellow going on the heart harvest.

Frank's every cell seemed charged with the conviction, the knowledge, that life was amazing. It was joy and it was madness, and sometimes, he knew, it was impossible to tell the difference between the two. This was the world he wanted to live in. He'd chosen at every step the hardest road available: the most work, the hardest challenges, the greatest responsibility. He couldn't imagine its being any other way. What was the point of all this, anyway? What was the point of being alive?

"Life style. Lots of people make life-style choices when they go through medical training."

We'd gone back to Frank's apartment, in an old brick building in an area of the city called the Flats, on the banks of the Cuyahoga River. He'd poured us a couple of glasses of a really good wine and put on some jazz while we waited for his girlfriend. He'd been in Cleveland for six months now, but he still seemed to be living out of boxes, and the walls of the warehouse-style condo were vast blank spaces.

"You go through med school and they say, 'Oh, don't be a surgeon—lousy life style.' It's a mantra in med school: 'Life style, life style, life style.' Do you go into emergency medicine, or do you go into . . . *life style*? I see people doing things that are really difficult, really hard and uncomfortable, but they do them anyway because they're passionate about their work. I want to do that. I see someone turn off his beeper because it's one minute after five o'clock—is that being a doctor? This is why surgery is the wrong choice for someone like that: no life style. That's what it comes down to—either you go with what you're passionate about, or you go with life style."

"So how did you first move toward surgery," I asked, "and why heart surgery and why pediatric heart surgery?"

"I wanted to do something with kids, so I did a general peds in med school. At the time I just thought I might want to explore pediatric surgery. But the chairman of the department wanted everyone to go into general pediatrics. I hung out with a great guy, Steve Black. He wanted to go into surgery, too. I got great reviews, and when the chairman asked me what I wanted to do, I said surgery. When he asked Steve, Steve told him he wanted to be a general pediatrician. Steve got an 'Excellent,' and I got a 'Satisfactory.' I went to complain, and the chairman said, 'I can give out only so many. Steve Black said he wanted to be a pediatrician.'

"I was stunned. That was it—and it was beautiful, it was clear: Screw the teachers, I'm going into surgery.

"When I was in school, I admired a guy named Paul Humphrey, who's now a vascular surgeon in Indiana. His father was a famous surgeon, too. Humphrey had the gift. Humphrey walked on water. You can see the guys as they go through; they weed themselves out. I wanted to be like that. I did some work, briefly, at Mass. General in Boston, and there was a resident there named Mark Moscovitch, a smart bastard. I admired him. I asked him what he was going to do, and he said, 'Heart surgeon.' You get these small impressions along the way—they kind of lead you—that the best go into difficult surgery. Neuro, vascular, plastic surgery—everyone will say theirs is the hardest, but one thing was always the case: the golden boys went into heart surgery. I saw people shake in front of the guys who were heart surgeons. And I wanted to be the guy who wasn't afraid of them. Same as my relationship with Roger: I will be so on the ball that he can't touch me.

"In med school the surgical world made sense to me. I wanted it to be the final answer. I didn't want to have to ask someone else to help me out during a trauma. I wanted to be a guy like Paul Humphrey or Moscovitch. I wanted to be the guy who was the calm in the storm. I wanted the art and the passion instead of the life style. I might have been a trauma surgeon. You can save lives there, you can be the calm in the storm, but the tragedy factor is big. Some people thrive on that, but for me too much tragedy is unhealthy. Whereas if you're a heart surgeon, you've got a problem, but it's a *scheduled* problem—you can think about it the night before, figure out how do I fix this problem? How do I fix it so that I find some art, some beauty in it, some kind of perfection for this baby? And then, *boom*—done.

"Pediatric plastic surgeons, they do some great work. Reid Hansen affected me a lot. When I was working with him at the University of Illinois in Peoria, he repaired a bad cleft lip. Really funny-looking kid, but Hansen helped to fix it—the kid had looked badly damaged, and now he was going to look more or less normal. We had just extubated this kid and he was doing well, and at the next bed there was a pediatrician checking the electrolyte balance of a kid who had rotavirus. We both watched the pediatrician for a minute, then looked down at the new beauty in the baby in front of us, and Hansen said, 'This is the best.' And for me, at that moment, it was very clear. Those moments tell you where to go.

"When I first worked with Buzz Helseth, my mentor, my senior partner, he was a very precise guy. I worked a year with him, 'ninety-four and 'ninety-five, scrubbed in on three hundred cases. He made a big impression. I liked precision, liked the honesty of it. For a patient, for anything, it wasn't a matter of being OK. 'How's the potassium?' Resident says, 'It's OK.' Buzz wanted to know exactly what it was—he didn't want a value judgment. 'The crit's OK? There *is* no OK—OK is *bad.*' *OK is not OK; we don't live for OK.* And if you're part of the population that believes in OK, then go with them. I don't want to be part of that ninety-eight percent."

"But there had to be a time," I said, "when you and all the rest of that two percent were *worse* than OK, when you were just learning. I know in general surgery they start with simpler cases like appendectomies, but how do you start on a *heart* when you're not very good, when you're new?"

"First thing you do is cannulate, put 'em on the heart-lung machine. And you screw it up, guaranteed. The first time you put those stitches into the aorta on an adult, you screw it up, and the guy training you says, 'What are you doing?!'—gives you a verbal smack. When you're in there, you know he's not just saying 'What are you doing?' He's saying, 'Jesus, you're hurting this guy,' and 'Watch it, man—if you tear that, this guy's leaving in a fucking box!' And that's when you learn that detail. You don't screw that up again.

"You do it twenty times, forty times, and then they let you sew a proximal coronary, a big vein the size of your pinky, into the aorta. The whole field is still. You're just getting used to the tissues. You have to sew it in a perfect circle. That's why some cardiac surgeons stay in adult, maybe: the beauty of the circle. You never get tired of it. And after you do that, then you go to the distal, tiny arteries, using thread smaller than your hair, seven-oh, eight-oh

Prolene. And that's the real thing. The key there is still to sew in a perfect circle. And the beauty is in the rhythm you develop. Four this way, *boom, boom, boom, boom,* turn eighty degrees, backhand four, *boom, boom, boom, boom,* and you're done. On a nearly microscopic level you develop a perfect pattern throwing a little vein onto a little artery. There's a beauty to sewing a perfect circle. You can say, 'I'll never achieve perfection, but I can come close,' and that's what makes a great surgeon.

"And then peds—yeah, it's the hardest. There's peds neurosurgery, too—those surgeons are a different cat, no doubt about it. They'll say their specialty requires a lot of three-dimensional thinking, but *our* three-dimensional thinking is instantaneously obvious when it's not good. Theirs isn't quite so immediate. If we do something wrong, the heart stops beating—*boom,* honesty. In neuro a kid comes back and there's a little weakness in his arm, or some residual tumor nobody found; it's not like suddenly there's no blood pressure.

"You cannot lie in this work. I love that. There are so many people in this world who have no idea who the fuck they are. When you do this work, your entire body is charged with it. You can be dishonest with *yourself*—maybe you don't know your limitations—but again, the results will speak for themselves. Maybe you become an asshole—you get bit, you get mean. Maybe you say, 'I don't have an ego.' That's another line. I mean, *I've* got an ego the size of a house. Steve Davis says that between us we have an ego bigger than Cleveland. You gotta have an ego. Whether it's the first time or the twentieth time, operating on a heart never loses its significance. When you're the surgeon, you feel it, and it is a *big . . . weight.* It's not the baby's weight. It's not two kilos. It's the mom, dad, grandparents, best friend, and cousins who live or die by Roger's hands or somebody else's. It's the weight of the entire community. The weight of your team and their expectations of you. It's the weight of your future, right? It's the weight of your legend, whatever you've constructed. Every single case is a test. You can't approach anything with a so-what—'Barney's operating, have a good day.' It doesn't work that way. Because I'm gonna judge you on that one. And the next one. And the next one. Forever.

"Every time you walk into that O.R., a thousand-pound creature squats down on your shoulders."

This was a young heart surgeon talking—thirty-five years old. He
had scarcely begun. He was already officially a partner, with his
own practice, at the Children's Heart Clinic in Minneapolis when
he came here to do this fellowship, but he was still relatively inex-
perienced. He had done no complex cases—switches, Norwoods,
truncus repairs—himself, only the simpler stuff: ASDs, VSDs,
coarctation of the aorta. This was the norm at his age; this was how
long the training took. Jonathan hadn't done even that much when
he first arrived, five years ago. It had taken him five years to get
enough experience under his belt to go off on his own. So what I
was seeing was Frank psyching himself up, after fourteen years of
training, for an intense career as a peds heart surgeon, maybe even
starting to construct his own legend.

"Right! *Heeeeee*. You got it, bro. Exactly!"

5. A Beautiful Heart

The *whoosh* of the automatic doors announces Frank's arrival in the O.R. He steps in wearing a hat, a mask, and surgeon's loupes, waving his arms gently to air-dry them before a nurse gowns him and a scrub nurse unwraps his gloves. "I just want to make sure that if we have to crash him on ECMO—" He pauses. "I figure if I scrub, we won't need it. And I think Roger can pull it off. I think we can make this kid better." Then, before he approaches the table, he turns back to me, his omnipresent grin visible behind his mask and glasses. "*And Alpha dog wants to open,*" he says.

Opening a chest a second time is especially risky; about 30 percent of the cases here are reoperations. Opening a fresh chest always carries some risk, too, but a chest that has been previously opened and has had time to heal will typically hide a mass of grizzled scar tissue, which will often have cemented itself to the underside of the sternum. When Frank or Mac yanks out the wires that were used to close the sternum and then passes a saw blade through it, he'll risk cutting into the heart or the aorta if one or the other has attached itself to the bone. And if that happens while he's opening, when they're still miles away from putting the patient on pump, the procedure may end up being a very short one indeed. So Frank, naturally, wants to be the one to open this chest.

The baby on the table is Drew H. The board gives the basic info—four months old, height fifty-three centimeters, weight five kilograms, blood type O positive—and a description of the operation: "Reop: MSOH: Revascularization of occluded LAD." The left anterior descending (LAD), the coronary that feeds the left ventricle, is blocked.

A little after one o'clock this afternoon, Mike Fackelmann, the O.R. P.A., anesthesiologist Paula Bokesch, and nurse Mark Myer entered room 1 of the PICU to prep Drew and take him to the O.R. The whole family was there surrounding Angie, who continued to agitate her baby—*still*. She'd been doing it since I got there before eight o'clock, shaking him and then shifting her hips to keep him from crashing to the floor as he writhed in her arms. Because the coronary that's supposed to feed his systemic ventricle is blocked off, that left ventricle has virtually stopped functioning; it's now ejecting only 8 percent of its volume. Drew's body isn't getting enough blood, and therefore oxygen, so his tissues are starving. As I watched Angie jolt her baby, it finally dawned on me: he was dying—this was what it looked like. Baby Drew, pale and skinny, was trying to die, and his mother was shaking him—shaking him *hard*—to stop him from slipping away. She hadn't been trying to calm him this whole time, or soothe him; she'd been fighting to keep him alive, the way you'd slap someone who wanted to fall asleep in the snow. He was—*is*—that close to death.

Paula, a lean woman with dark hair who was a fellow in Boston in the early 1980s, when Bill Norwood was developing his procedure, took Drew out of his mom's arms. Angie breathed hard and sniffed back more tears. Paula knew how sick this kid was, knew he could arrest at any moment, so she focused all her attention on getting him into the warmer bed and then, with Fackelmann's help, to the O.R., without losing any of the lines that were delivering drugs to him. She was so intent on him, in fact, that she all but ignored Angie, as did Mike. Mark, the nurse, was gathering Drew's records and a fat green three-ring notebook from a rolling tray when he noticed how scared Angie was. He took a step toward her and promised, "We'll take real good care of your baby. He's got an excellent surgeon. There'll be nurses to let you know how it's going."

Angie nodded. Just twenty-five years old, she was gray and exhausted, and she had once again given up her baby boy, her only child, to strangers. This time, the purpose of the operation was not simply to fix him but to try to salvage his damaged heart and thus save his little life; it was his last chance. It was almost too much for her to bear, and she might not have been able to get through it had she not felt that Drew somehow understood.

About ten minutes before Mike, Paula, and Mark came to get her four-month-old baby and take him to the operating room, Angie had said a prayer and spoken to him. Her parents were with her when a nurse told them that the O.R. staff would be down soon. She stopped shaking Drew and laid him in the warmer bed for a moment, then leaned over him and said softly, "You're going through a major surgery again. Mommy needs you to be strong. I need you to be tough. I need you to follow the Lord—he's going to lead you to where you need to go." And Drew quieted for the first time all day; he stopped writhing, and his helpless cries ceased. His vivid blue eyes looked straight into hers. "Mommy loves you," Angie said, "and you just need to be strong and fight." His eyes still locked on hers, Drew slowly lifted his arm and touched her face. His grandmother, amazed by this little exchange, gasped, "Angie, he's *listening* to you." Angie, for her part, believed that her son was telling her he would do his best.

Angie and Bart have been married for a little over a year. Born in neighboring towns, they played together as preschoolers because their parents were friends, and even though they grew up in different school districts, they remained peripherally aware of each other. After graduation Angie married her high school sweetheart, whose military service took the couple to Kansas. Angie is direct and straightforward about that chapter of her life. "Kansas sucked," she says. "You know how they say there's nothing in Kansas? There's nothing in Kansas." The marriage ended quickly, and Angie returned to live with her parents and start over. She found work as a medical assistant for some family practitioners in private practice; it was her job to escort patients into examining rooms, record their height and weight, and take their blood pressure. She hoped to pursue nursing as a profession. Both her parents and Bart's spent a lot of time at the American Legion hall in her hometown, playing eu-chre. Bart occasionally joined them, though he didn't much care for the game. A quiet, sturdy man a year younger than Angie, he'd gone straight into the family fence business after high school. He'd gotten a football scholarship to a small college in Delaware, but after sustaining a back injury in a high school game, he'd decided to decline it, mostly because he just wasn't that interested in going to college. He was happier working, and when he wasn't working, he loved to watch professional motor sports—he's fascinated by "any-thing with an engine," he says—and go hunting. But it was during a game of euchre at the lodge that Angie's father put it to Bart: "So, are you ever gonna ask my daughter out?"

This struck Bart as odd, because he and Angie were so unthink-ingly connected that they sometimes called each other Cuz. But one day he did ask her out. His last girlfriend, a longtime high school steady, had been stylish and showy and expected him always to act the gentleman. And so Bart did just that with Angie, figuring it was what all girls wanted. He took her to a steakhouse, smoothed a napkin over his lap, and made polite conversation. He ordered a soda pop because he was driving. Angie was about ready to die of boredom. She asked the waiter for a beer, and after taking a few good swallows, she opened her mouth and lofted a guttural belch across the table. Angie now closes her eyes in embarrassment when Bart tells the story, but the tactic worked: Bart took the napkin off his lap, picked up the steak bone in his hands, and chewed on it, as he was inclined to do. Angie was something altogether different from what he'd known. Within six months they were engaged.

About two months before their first wedding anniversary, on July 18, 2000, their son, Andrew David, was born. He weighed seven pounds seven ounces; the delivery was normal, and all seemed well, just as all was well for Bart's sister, who had given birth to a child of her own a few days earlier. The next morning a nurse came to take Drew to the nursery, where he'd be seen by a pediatrician. Angie kissed her son and told the nurse she was going to get cleaned up and ready for visitors eager to see the baby. But within ten minutes the nurse had returned to help Angie get dressed, explaining that the doctor needed to talk to her about the baby's checkup.

As the nurse was wheeling him to the nursery, Drew had begun to cry and could not be comforted. The nurse noticed then, as he continued to scream, that he was starting to turn blue. She ran for the doctor: the baby had stopped breathing. He was immediately resuscitated and taken to the ICU.

"There is something very wrong with your little boy," said the doctor, waiting to greet Angie as she emerged from the bathroom. Those words still ring in her ears.

What had happened with Drew was not atypical of babies born with TGA, transposition of the great arteries. They're fine for a while, so long as the ductus stays open—the ductus being the patent ductus arteriosus, or PDA, which maintains communication between the aorta and the pulmonary artery. But then, as tissue seals up the openings in both arteries and the duct itself disintegrates, the reversed arterial scheme begins to reveal its perverse mission, circulating depleted blood continuously through the body, and oxygen-rich blood through the lungs. Unless there's a septal defect—about 60 percent of babies born with TGA also have a ventricular septal defect (VSD), a hole between the ventricles—the blood can't mix, and slowly *all* the blood in the body begins to turn blue. And so it was that Drew, whose duct was slowly shutting down on schedule, gradually developed a distinct blue cast.

The urgent—but standard—procedure for such cases quickly went into effect: doctors at the community hospital started Drew on prostaglandins to keep the duct open and transported him by ambulance to the nearest facility that could identify and treat congenital heart defects, a larger children's hospital. Angie and Bart followed in their car. At the children's hospital a cardiologist diagnosed TGA/IVS (transposition with intact ventricular septum) and performed a balloon atrial septostomy, threading a tube with a balloon on the end into a vein in Drew's groin, then up into his atria, and popping a hole between them to enable better mixing of the blood. The baby boy was then allowed to stabilize and grow for a little less than a week in intensive care. The chief heart surgeon—I'll call him Dr. Jones—told Angie and Bart that there was a 5 percent risk with the switch operation, and that one of the biggest potential dangers was infection. Dr. Jones never used the word *death*, according to Angie, and she herself didn't think to equate "risk" with mortality. She and her husband knew the situation was serious, but they were not inordinately worried that their baby would die. Drew went in for surgery on the morning of July 25 as scheduled. Between two and three hours later a nurse reported to Angie and Bart that the surgeon had found an abnormality called an intramural coronary, which was unusual in this particular form of transposition and potentially a problem. Angie recalls the nurse's telling her that 5 percent of patients with intramural coronaries didn't leave the operating room alive.

No pediatric heart surgeon will deny that the discovery of an intramural coronary makes for a more difficult switch. A normal coronary enters the aorta directly, like a garden hose screwed straight into the side of a bucket. An intramural coronary, by contrast, is like a hose that enters a bucket, winds along inside the wall, and then emerges some distance away from its point of entry. Transplanting such a coronary into the pulmonary trunk (or neo-aorta) requires some creative improvisation and the skillful use of flaps and hoods to avoid cutting off the blood supply. There are many distinct types of abnormal coronaries—sometimes, for example, there is only a single, left coronary that branches left and right, and sometimes both coronaries are intramural, and while many of these have been categorized, each pattern is unique within the entire scheme of the individual heart, where the sizes of the aorta and the pulmonary artery and their spatial relation to each other are always different, rendering the total number of possible variations infinite.

After several more hours Angie and Bart were given the bad news: their little baby had been unable to come off pump, so his chest had been left open and he'd been put on ECMO (the extracorporeal membrane oxygenator), a last-ditch device that within a matter of days could cause multiple organ failures or massive damage to his central nervous system—if, that was, ongoing heart failure or sepsis didn't kill him first. For those patients who can't be weaned from the machine, an aggressive move toward transplant becomes the only hope of survival. Mortality for children on ECMO ranges from 30 percent at best to almost 80 percent, so it's always a race against the clock.

Then things got even worse. Drew's cardiologist—Dr. Murray, I'll call him—performed a catheterization to try to learn why Drew was having such a hard time of it, whether the underlying cause was pulmonary hypertension or a coronary issue. During the cath lab procedure, as Murray was feeding the catheter into the aorta and up the aortic arch, it snagged on the ECMO cannula, and the baby began to bleed. There was, Dr. Murray said, "too much hardware in there" for such a small heart. Angie and Bart were immediately informed of the problem and told that their son might not survive the procedure. They went to the hospital chapel to pray. A surgeon was paged—Dr. Jones's partner—and he opened the patched chest, found the bleeding vessel, and closed it with a stitch. Angie and Bart emerged from the chapel and were met by a friend of Bart's who'd come running to tell them the news: the problem had been fixed. Drew would make it out of the cath lab.

The cath results showed delayed filling of both coronaries as well as some blockage in the left one. "We were concerned, but we didn't think he needed to be reoperated on because there was some flow," Dr. Murray recalled when I called to ask him about Drew's case. The high dilation pressures also suggested that there might have been some insult to the left ventricle from the surgery or the cardiopulmonary bypass, but the baby's surgeon and doctor hoped that with time and medicine, his heart function would improve.

Within days the doctors began to wean Drew from ECMO, and happily, he appeared to tolerate it. Eventually they got him off the machine completely and were able to close his chest.

Angie would sit with her son while he was still on ECMO, his chest opened, a football-shaped sterile patch sewn over the opening, and what seemed like an enormous number of tubes coming into and going out of his body. Her friends asked how she could bear to see her baby in such a horrifying state, how she could bear what he'd been through. She told them, "I don't even think about that. I don't walk in and look at all the machines and wonder,

What's this and where's that going? I just walk in and see my baby and look at him and see how beautiful he is and pray that he's getting better."

And indeed, he seemed to be doing just that. By mid-August Drew's doctors felt he was well enough to go home on Lasix (a diuretic that reduces swelling and edema) and aspirin, which acts as a mild blood thinner. Murray scheduled a series of weekly checkups for the baby, but after just a couple of these, he said that Drew seemed to be doing fine and that Angie would not need to bring him in again until January.

But he was never really very well, Angie says now. It was obvious that *something* was wrong. Of course, they couldn't expect him to be completely healthy, given all he'd been through—and how were they to know what was normal or not for an infant after major surgery, two caths, and ECMO?—but even so, Drew looked awfully sickly, and he didn't appear to be growing much. Angie and Bart rent a small, comfortable house on the edge of eight hundred acres of corn and bean fields—the house Bart grew up in, surrounded by the semirural streets he biked on as a kid—that they hope one day to buy from the woman down the road who owns it. From the day they brought the baby home, they slept with him every night, perpetually worried; the doctors had told them not to let him cry, if possible. Otherwise they carried on as a small family as well as they could, until Drew simply stopped waking up to eat. When he *did* eat, it was painful to watch: he couldn't take more than four or five sucks from a bottle before quitting in exhaustion, sweating heavily. This wasn't right. Angie and Bart took him back to the hospital in mid-October, six weeks after their last visit. Dr. Murray started Drew on dobutamine, a potent inotrope, and did an echocardiogram that revealed what was causing his difficulties: his left ventricle was dilated, meaning that it was filling up beyond its capacity to contract—it was failing. A catheterization performed the following week confirmed that the left coronary, the one that Dr. Jones had originally identified as being intramural, was blocked. "Drew didn't get better like we thought he would," Murray conceded.

Dr. Murray explained to Angie that either the coronary had become blocked off because of scar tissue or some sort of clot, or it was kinked. Whatever the reason, however, it was a problem they didn't feel comfortable trying to correct surgically. He and his colleagues had discussed Drew's case in conference and had decided to proceed with a medical therapy: they would give him powerful doses of dobutamine and wait to see if his heart would develop collateral arteries from the right that would feed the left ventricle. "You always hope you can put somebody on medicines"—as opposed to having to operate—Murray told me. Angie and Bart did as they were advised, returning home on October 21. Three days later, though, Drew looked terrible again, and he was vomiting frequently. Angie took him back to the hospital, where his medications were increased. Over the following week the baby showed no improvement, only deterioration. Dr. Murray acknowledged that the medical therapy wasn't working but said there was nothing more he could do for Drew. There were two options left: either Angie and Bart could take their son home, care for him as best they could, and hope for the best, or Murray could contact Roger Mee in Cleveland and ask if *he* could help Drew somehow. Dr. Mee, Murray predicted, would respond in one of three ways: he would either confirm that nothing could be done, offer to attempt a coronary repair, or recommend a transplant. Angie asked Dr. Murray to write to Dr. Mee, which he did on November 2. His letter posed a series of questions: Might Drew be a candidate for revascularization, an operation to fix the blocked coronary? If so, would Mee himself be willing to do the surgery? And if such surgery *wasn't* a viable possibility, should the baby be evaluated for cardiac transplantation? Murray noted in conclusion that the patient "is stable on his medications and does not have the appearance of an infant who will have a further unexpected fatal cardiac event."

On November 8, after reviewing the O.R. notes, the catheterization report, and the angiogram, Mee sent Murray his assessment of the situation, which concluded, "Although it would be difficult, I do think that this patient should have the potential benefit of revascularization." The letter closed with thanks to Dr. Murray for seeking his opinion, kind regards, and yours sincerely.

Upon hearing this news from Dr. Murray, Angie and Bart met

with the rest of their extended family at Angie's mom's house to talk over the situation. They all agreed: they would try to do everything within their means to save Drew. Angie spoke with Debbie Gilchrist in Dr. Mee's office the following Monday. The surgery was scheduled for the week after that, Wednesday, November 22; Debbie told Angie the family should plan on arriving Tuesday morning so the baby could be admitted and the staff could get started on his workup, which would take most of the day. Angie and Bart accordingly prepared to leave for Cleveland Monday night so they could have Drew at the hospital first thing Tuesday. On Sunday night, as was their routine, Angie and Bart left Drew at home with Bart's mom's sister, Aunt Ruth, and went bowling. (They were rarely away from their son at all, and when they were, he was always in the care of either his great-aunt or his grandmother, Angie's mom.) Their evening out was cut short, however, by a call from Ruth: the baby had turned an awful color and was throwing up. They returned home immediately and phoned Dr. Murray, who met them at the hospital. Drew was put on a powerful infusion of dobutamine, and kept in the ICU overnight; arrangements were made for an ambulance to take him to Cleveland the next morning, a day early. Angie went with him, while Bart and his parents followed in the truck that afternoon, as the snow began.

The next morning was when I first saw the family, saw Angie shaking her sick child to keep him alive. After we left Drew's room on rounds, I spoke with Dan Murphy, the center's head echo man. At age forty-eight, Dan had already been practicing for sixteen years, having started when echocardiograms could provide only one-dimensional pictures of the heart. (Larry Latson, who heads the cardiology department at the center, likens reading those early heart echoes to trying to describe a glass elephant in a dark room using a penlight.) He had worked in Texas with Denton Cooley, one of the country's pioneering cardiac surgeons, and had been at the Cleveland Clinic since 1989. Referring to Drew's case, he said, "There are probably half a dozen people in the world who can handle this, and Roger's one of them." When surgeons first started performing switches, Dan told me, they'd needed angiograms to determine whether the coronaries were complex. Mee had been one of the first to dispense with that tool—he'd simply started going in and figuring out how to fix whatever he happened to find. "Roger's a magician with those," Dan added. "But there are two parts to the magic: that Roger can deliver blood to the heart, and that the heart muscle can come back."

In the O.R., the baby squirms so violently in Paula Bokesch's arms that she can't both carry him and administer the ketamine that will put him to sleep. "Here, hold him," she says to Carlos Gomez, an anesthesiology fellow. Carlos holds Drew awkwardly, but he's struck when he sees the boy's face—"Look at those eyes," he says. Paula remains tense. Earlier, Roger was so nervous about the case that he called her to discuss intubation. The problem is, anesthetizing and intubating a baby who's this close to the edge—"circling the drain, to use a medical term," Paula says—could actually kill him by depressing his already compromised heart. Then, too, his body has likely adjusted its physiology by altering its levels of epinephrine and other catecholamines—chemicals that set off the fight-or-flight reaction that helps us stay alive in dangerous situations—but that balance can be precarious; at any second he could have a heart attack right there in Carlos's arms. Paula delivers the ketamine through an intravenous line. When the baby is asleep, thanks to the drug that will also maintain his blood pressure and heart rate, she paralyzes him with Pavulon. She keeps a close watch on his pressures and heart rate, but the process goes smoothly. Together she and Carlos lay the baby on the table, where she will intubate him. She cranks his tiny head back and slips a bright silver intubating blade into his mouth and down his throat, peering down it as she goes to make sure she's past the vocal cords and into the trachea and not in the esophagus—she's got to be fast and accurate here because the baby's lungs are paralyzed, and he has no way of getting oxygen until she has the tube in. Carlos sets up the drips while Mike Fackelmann retrieves a Foley catheter. Frank Moga asks Kevin Baird, the perfusionist, "Do we have an ECMO ready?"

Kevin says, "We have one dry that I can hook up pretty quick."

"I can drape him while somebody does CPR."

After getting the Foley in, Fackelmann uses a sterile black marker to draw bold X's on each groin, marking the spots where his fingers tell him he can get access—a concern, given that these vessels may be thrombosed, or clotted off, because of all the procedures Drew's been through. Paula feeds an arterial line into the right radial artery. Next she swabs the baby's neck with antiseptic, just over the right jugular, so she can insert a double-lumen catheter down this vein, into the superior vena cava, and into the right atrium of the heart, where it will monitor pressures. Frank steps up and regards the infant, who is splayed naked on the table. His chest is lumpy, the left side slightly higher than the right; Frank says to himself, "Attention to detail."

Prepping continues routinely, with Kevin readying the bypass machine, scrub nurse Bob Cherpak organizing the setup table, and Paula and Carlos working on the lines. Fackelmann sponges Drew's entire upper body with Betadine, an inky brown antiseptic, then stretches a yellow adhesive sheet over his chest; this Betadine-infused draping will supply further protection against infection. Frank and Fackelmann now begin to drape. The cage is locked into the table above Drew's head and draped as well.

Dan Murphy comes in wearing one of the spare gowns that usually hang by the O.R. doors, a bouffant cap, a mask, and paper booties over his penny loafers. He's been paged to do a TEE, or transesophageal echo. A long black tube with a handle and a cord on one end has been inserted down Drew's esophagus. Dan wheels the echo machine to the head of the table and plugs the cord into it. Because the heart lies right on top of the esophagus, a TEE will provide better pictures than a regular echo, where a probe would be passed over the surface of the baby's chest and belly.

With his back to the table and one arm reaching back to manipulate the probe, Dan faces the screen and gets clear pictures of Drew's heart. He sees an active right ventricle but almost no activity at all in the left, only a weak undulation instead of a contraction. "We used to think that was all dead muscle, but you never know," he tells me. The muscle could simply be "hibernating." After taking various measurements of the valves and flow, he writes his eval-

uation on a piece of paper headed "Echo Report": "Massively dilated LV and severe global LV dysfunction."

The head anesthesiologist, Emad Mossad, forty years old and a
native of Cairo, Egypt, drops by the O.R. to check on the patient
he helped admit. "He looked dead yesterday," Emad says, glad that
the baby has made it to the O.R. and been successfully intubated.

"Starting skin," Frank announces.

"First incision," says Debbie Seim, the circulating nurse, as she
notes the time: 1438.

Frank drags a scalpel along the bumpy ridge that marks Drew's
sternum. The actual reopening, a procedure that can be deadly if
the aorta or the ventricle is stuck to the sternum, is completed
without incident, slowly and cleanly. The sternum is then ratcheted
open to reveal a dark, gristly pericardium, not an unusual sight in a
redo. A baby's heart is normally smooth; it shines. The colors are
soft pinks and reds. The two main vessels rising out of it are clear
and distinct. The coronary arteries create a well-defined vine of vessels running over the myocardium. Drew's heart, in contrast, is
now a dark, ugly red; it looks almost as if it were covered with bar-
nacles, from the scar tissue that has grown all over it. Frank must
take down and cauterize numerous vascular pericardial adhesions,
which sizzle and turn black, sending smoke rising out of the open
chest. This heart is also strangely large. The patient is on his back,
so the right atrium and right ventricle are facing up; these chambers
are what Frank and Mike see, and they're beating well, as Dan's
echo showed. Not visible on the other side is the swollen, useless
mass of the left ventricle.

Roger Mee enters the O.R. uncharacteristically early. In difficult cases he is actually afraid, and the fear is specific: "fear that I
might hurt somebody," he says. In this particular instance it seems
a little more serious than usual; he has wondered aloud to me if he
hasn't brought the baby here only to "assassinate" him. Fear is
good, he believes, if one can use it properly. Some people become
paralyzed by it, and that's bad, but fear can also be used to advantage and is therefore desirable. He loves, he says, "straightforward
cases that go just like poetry"—cases about which he feels no fear.

When he is genuinely afraid of hurting a patient, he says, "That's not relaxing." For him fear is a tool: "I use it to drive myself." The fear factor is higher in Drew's case because there's a possibility that he'll die, but there's an even greater possibility of that if Roger does nothing. The only conceivable chance this baby has, aside from the operation Mee is about to attempt, is a transplant. But the odds that a suitable donor heart could be found in time are slim to none, given how sick he is. To understand just how long it can take to get a heart for a baby, one need only look into the room of Ashley Hohman (an infant of approximately the same size as Drew), where Kelly, Ashley's mom, is still doing her word puzzles, nuzzling her daughter's cheek, or staring at the monitor—seven weeks on the transplant list, and nothing yet. This operation is Drew's only realistic chance, Angie and Bart's only chance.

"Are you there yet?" Mee asks Frank.

He is. "It's a strange new world," he responds.

"Do you want a full dose of phenoxy?" Paula asks.

"Yes, I think so," Roger says.

When she asks how much blood he wants available, Mee's answer is, "I think we're going to be here a long time." She nods and hangs the first bag of irradiated O positive blood on the rolling metal pole, and Roger asks Bob Cherpak if he's got the 8-0 Prolene stitches ready. Stitches of this size are so fine as to be almost invisible; though Roger rarely requests them, he believes it's critical to use the smallest possible material on a micro bypass that won't tear the vessels. After Bob confirms that he's got them, Roger leaves the O.R.

If the floor of the O.R. were glass, we'd be able to look down now and see a big group of people in the parents' lounge below: not only the family of the child Jonathan is operating on next door, in the second O.R.—he's just finishing up the repair of a lesion called complete atrioventricular canal in the heart of a five-month-old baby with Down's syndrome—but also thirteen members of Drew's extended clan, who have gathered here in this hospital, a few days before Thanksgiving, to share the burden of waiting, to add the weight of their prayers to all the prayers that have already been said for Drew. This surgery is their last hope. In addition to Angie and Bart there are Angie's mom and dad, Bart's mom and stepdad and his dad and stepmom, his younger brother Seth, and his uncle Bart and aunt Ruth, uncle Jim and aunt Pam. Angie's brother isn't here because he had to stay at home with his daughter, Drew's senior by only a few days, but there's a phone in the lounge that the others will use to call him frequently throughout the afternoon. When he gets a call, he'll then phone all the great-grandparents. And he'll also call the church. Angie and Bart's church has put up a picture of Drew, along with an explanation of what's happening today and a plea to all to pray for the little boy. The church has also contacted other churches and asked them to do the same, to ask *their* parishioners for prayers, and they've called friends of theirs, other churches, asking them to spread the word. Angie has been told that this prayer chain stretches as far south as Georgia: complete strangers across the country, all praying for her baby.

The parents' lounge is comfortable but impersonal. Here Drew's family will wait and wait and wait. They will watch TV. They will play cards. They will look up at the ceiling frequently, knowing that their youngest member is immediately above their heads, and say another prayer, or even say out loud, "Hang in there, Drew."

It's been snowing since yesterday, and now the snow has picked up. Already the two main avenues leading east out of downtown Cleveland, Carnegie and Euclid, are filled with traffic, though it's just three o'clock. Many downtown workers, apparently concerned about getting home, have left work early. The streets are white, and the snow is expected to intensify.

Mee reenters the O.R., plugs his headlight into the generator, adjusts its light on his palm, then leaves again to scrub. "When Roger's in the room," Frank told me earlier, "everybody's sphincter is a little tighter." Given the nature of this case, he added, he expects the effect to be doubled today. Mee returns to the O.R. with his hands and powerful forearms dripping. He dries them with a sterile towel, unfolds a gown, and slides both arms through; the circulating nurse, Debbie Seim, ties it behind him. Bob the scrub nurse removes a pair of gloves from their paper wrapper and holds them out for Roger, who looks, in his hood, headlight, louped glasses, and gown, almost like an underwater explorer. Frank steps down and away from the table and moves carefully to the opposite side, to the patient's left. Debbie plugs in Roger's headlight, and he steps up to the table. He picks up forceps to get his first look at this heart: "Holy shit," he says, the words soft and obscured behind his mask. The heart is so big, so swollen, that gaining access to the switched vessels won't be easy. He sees multiple arteries leading from the pericardium into the muscle of the left ventricle, the result of efforts by Drew's body to grow vessels that might feed the dying myocardium.

Four people are scrubbed at the table besides Roger. To his immediate left, in the "monster boy" spot, is a second international surgical fellow, whom I'll call Charlie. He is thirty-six years old and halfway through a yearlong fellowship here. At about five feet three, he has to stand on three footstools in this position. Roger has never before encountered a situation in which height so adversely affected a surgeon's ability to assist. To further complicate things, Charlie is tentative in his movements and hindered somewhat by his imperfect English. Bob Cherpak is on Roger's right, from which place he will be able to slap tools into the surgeon's hands and squirt water on the heart or on fingers holding sutures to keep them slippery. Across from Bob are Mike Fackelmann and Frank Moga. Behind the cage, Paula Bokesch is overseeing anesthesia along with Julie Tome, who will take over for her during this long procedure.

Before Roger begins, he first locates a mammary artery high up in the left side of the patient's chest so if he needs it later—as he suspects he may—he won't have to use up precious bypass time searching for it. (Because cardiopulmonary bypass is a tremendous insult to the body's blood and organs, the less time a patient spends on it, the better.) He then gets to work putting Drew on bypass. In the dark-red ball of gristle that is this baby's heart, he finds enough ascending aorta to cannulate: a tube inserted here will shoot oxygen-rich blood from the machine behind Frank, manned by Kevin Baird, into the aorta to perfuse the brain, arms, and body as well as the coronaries. The heart will continue to beat, but it won't have much blood in it. Roger identifies enough of the right atrium to accommodate a single venous cannula, a tube that will suck in all the blood returning from the body and send it to the bypass machine. All of this is typically routine work, but today, perhaps because this case promises to be particularly tricky, the mood is tense. Charlie seems more nervous than usual, and Roger is quick to jump on him: "Charlie, you're dropping the ball. Either retract or get out of my way." Paula has some difficulty ventilating—there's a kink in the tube. Roger asks her what the blood pressure is.

"Forty-two, forty-three mean," she replies.

Moments later the baby's heart begins to race, almost doubling its rate, to 212 beats per minute, and his pressure climbs into the fifties. This is exactly what Roger doesn't want to see: the heart overworking before it's even on pump. Soon it will begin to fibrillate.

"Have a shock ready," Roger tells Debbie. She rushes to the generator as Charlie fumbles with the paddles; they're tangled, and a large steel disc is screwed into each shocker. "We need a small one and a big one," Roger says. Charlie swiftly sorts them out, then removes one large disc and screws in a small one, about the size of a concave nickel, as Roger says, "Good, untangle it," and then takes the paddles from Charlie, places one behind the heart and one on top, and says, "Hit it." A shock is delivered, and Roger says "Good" as the heart resumes its normal rhythm.

When he nearly has the cannula in, he says, "Heparin, please." Heparin is an anticlotting drug without which the baby's blood would immediately gum up in the machine and within his vessels.

Julie says, "Six hundred units, Kevin," as she depresses the plunger of a small syringe that's screwed into the central line.

"Six hundred units," Kevin repeats.

"Suckers on."

"Suckers on," Kevin says. Now that the heparin has been administered, the suckers can pick up blood and send it back to the heart-lung machine so as little as possible will be wasted.

Before bypass is begun, the pump circulates the blood through one long loop of clear plastic tubing. Then Roger says, "Pump off, ready to divide."

"Pump off," confirms Kevin.

Roger clamps the tubing in two places and cuts the tube between them. One end will be attached to the aortic cannula, the other to the venous cannula. This procedure, too, is routine, but the tense mood persists because Drew's heart is still beating too fast, working too hard. It fibrillates again, and amid the increasing stress, Roger says, "We're all right, don't panic. Hold that, Mike." He hooks up the tube to the aortic cannula and soon says, "Pump on . . . full flow." The heart empties of blood, remaining dark red but becoming flabby. Roger shakes his head and says, "Bad start, bad start." Kevin begins to lower the temperature of the blood to 22 degrees Celsius (71°F), to cool the patient in preparation for cross-clamping of the aorta and the delivery of potassium into the coronaries to stop the heart. Once again the heart begins to fibrillate, and to distend. "Right angle, please," Roger says. "Put a plege stitch in." Then he shakes his head and mutters, "Here we are, rushing to get cross clamp on."

"Pleging," Kevin says as he delivers the poison that will stop the patient's heart.

"What's your plege pressure?" Roger asks.

"I've got about thirty," he answers.

"Come up, it's a big heart."

As he's working, Roger nicks a small vessel, causing blood to well up in the chest cavity and swamp the heart. It takes him a full minute to sew it up. "Bad start," he continues to mutter. "Really bad start."

Roger Mee has a recurring nightmare, or variations of the same dream. He's in the operating room in the middle of a case; it's a difficult one, and he can't seem to finish it. He gets angry at his team. One by one they begin to leave; his assistants put down their tools and walk out, and he has to take over each person's job as he or she departs. Soon he's dashing around the O.R. trying to run perfusion and anesthesia *and* do the operation, all by himself. Sometimes it begins when he asks for the Prolene sutures and is told that the previous surgeon always used rope. Roger responds, "Well, I use Prolene." An assistant grudgingly goes off to look for some Prolene and never returns. And it just gets worse from there. One by one people disappear as Roger grows angrier. The perfusionist is needed elsewhere and rolls the bypass machine away. Eventually everyone else is gone, and Roger still can't finish the case. The dream ends only when he wakes himself out of it. In the most terrifying of the variations he finds himself trying to operate completely alone in the middle of a field.

The dream comes less frequently now, but he's had it often enough that the last time it happened, he was able to tell himself in the dream that it was just a dream. He tends to dream it, he says, not when he's worried about a case, not when the ICU is loaded with really sick kids, but rather when he's frustrated and angry over quality control in his center—by which he means when his people aren't doing their jobs well enough. The system is too intricate to allow him to monitor all its parts, whether the problem is a nurse who's taped a line in insecurely, a fellow who has inserted a chest tube slightly out of place, or a resident who has closed an incision awkwardly. He's hard on his staff and doesn't hesitate to give them a bruising. He feels he *needs* to be as tough on his team as he can without their becoming demoralized and simply leaving. It's a frequent question for him: "How hard can I push people?" he asks. "Because I need them. I absolutely need them."

Although nervousness or anxiety often seems to be an element of his mood, he's seldom outwardly nervous; it's just that at these times he appears to be more aware of all the things that could possibly go wrong. Some surgeons he knows, by contrast, out and out *enjoy* the O.R. Jonathan is one of them, and Nancy Poirier, who recently completed her fellowship at the clinic and is now practicing in Montreal—a rare female in this surgical specialty—is another. Then there's his old pal from residency days, Stephen Van Devanter, also a peds heart surgeon. "He loved to cut," Roger says of Van Devanter. "He *loved* it." Roger recalls the night a guy with a knife wound was being brought to the Brigham—they didn't get many of those there, he says—and Van Devanter was ready at the emergency entrance with scalpel in hand, waiting for the ambulance. The patient was reportedly dead, but Van Devanter cut him open on the stretcher as he was being carried out of the ambulance, got his finger on the hole in his heart, and, with Roger's help, saved him.

Roger's personality is the precise opposite. He openly tells parents—especially those who seem skeptical of surgeons' motives— "I don't love to dig around in children's chests." He says he's on salary at the clinic and therefore has no monetary interest in any one particular procedure, and besides, frankly, he'd prefer to be home with his family rather than in the O.R. He tells them, "I have the ability to do this if you want to make use of it." But ultimately he seems to do this work in spite of his own desires, and his greatest frustrations consist not in the big, complex cases but rather in the small details of "quality control" that are outside his control.

Roger has experienced awful conditions firsthand, situations that in the United States would be considered almost criminal. In the 1980s he traveled widely in China, operating at the behest of various hospitals to demonstrate his techniques to repair children's hearts. But operating through a translator proved difficult, because whenever things got tense the translator would forget to keep translating; then no one else would know what to do, and the operation would stall. The materials available were deplorable: the hospitals reused gloves, for instance, after patching any holes. "They were so poor they had to save things," Roger explains. In opening one chest, he recalls, he slid his finger under the sternum to ensure that nothing was sticking to it before he ran a saw through it, and when he pulled out his finger, it was bare—the glove had torn. During that same operation he touched the patient's aorta and felt that it was slack. Surprised, he asked the perfusionist if the patient was on pump; the perfusionist said he didn't know. On another occasion, scheduled to perform surgery in an O.R. crowded with Chinese observers, he got a patient on pump, got the cross clamps on, and stopped the heart with cardioplegia. Then he reached up to adjust the light so he could see horizontally inside the heart, only to discover that the lamp was fixed in place. He couldn't adjust it down and therefore wouldn't be able to see through the atrium into the ventricle and patch the hole there—it would've meant operating blind. "I can't operate like this," he said. "Quickly, I need a light." They would look for one, he was told. The patient was already on bypass, with cross clamp on and heart stopped, so Roger was losing valuable time and endangering a young life. "Anything," he pleaded. "A flashlight." He stood there waiting for twenty-five minutes, "about to shit my pants," he says. "I'll have to shunt this kid," he thought, instead of doing the repair, "and even that's not going to be easy without a light." At last two men entered the O.R. carrying a car headlight, wired to a car battery, the light itself lashed to a pole. The heat from it burned Roger's ear throughout the procedure, but the kid did OK.

Here in the United States, quality control is more about refining and perfecting details than jury-rigging lights or maintaining sterile setups. Roger is intolerant of imperfection, whether it's a tube running out of a patient's chest and over the edge of the ICU bed rather than over the foot of the bed to drain, a wall-mounted hand-sanitation dispenser that doesn't work, an assistant's poor retracting skills, or his own stitching. But this intolerance may just be what lies behind Roger's success—not just his low mortality rate, but how quickly his patients heal, as gauged by how soon after surgery they leave the ICU. Perfect stitching requires an intuitive sense of the importance of absolute accuracy. It's not difficult to understand: good stitching, for Roger, is common sense. "When you rivet plates together," he says, "you have absolutely evenly spaced rivets, so that each of them takes an equal amount of force." The same principle applies to stitching tissue. He draws a horizontal line on a piece of paper, suggesting two edges that have come together to be joined. "Your first stitch goes there," he says, putting a pen point a quarter of an inch above the line. "Once you've done that, you've predicated where every other needle has to go." He makes a series of dots, each one a quarter inch from the next and a quarter inch above the line. "You're really looking for squares, for every needle hole all along the way," he explains. If any one stitch is off even slightly—a little more than the quarter-inch predication, a little less—then the pressure on that point will be uneven, and tearing of the very delicate tissue will be more likely, resulting in leaks or a buildup of scar tissue. If the stitches are too close together, or if the surgeon pulls them too tight, the tissue will be strangled and will die. Perfect stitching is especially critical in heart surgery because the vessels have blood running through them at high pressures.

"We have a huge advantage over gut surgery, say," he notes, echoing a point he's made before. "If we misjudge the tensile strength of a vessel we've sewn together, we have a problem right there smack in front of our eyes—immediate feedback. Now, if you do that with the gut, not putting it under tension but maybe strangling it, it's five days later when the gut falls apart and you've got peritonitis, by which time you've forgotten where all the stitches were. This is why, in general surgery, there's this whole mystical thing about patients who are good healers and patients who are poor healers, and, 'Well, he had some congenital heart disease and his oxygen saturation was only ninety percent so he didn't heal.' Well, bullshit. Our kids heal with saturations of seventy percent. But we were lucky: we learned very, very quickly where to put the stitches and how tight they had to be. Plus we're sewing things that have a fairly high pressure in them. So if you've selected the wrong place to put your needles, and you've had to pull too tight, there'll be little tears, it's like a watering can. So you learn pretty quick, or you should do.

"If we were dealing with a mechanical system, there would be only one optimal way to do these things—where you put the screws and how tight. It's known and reproducible. What was blurred for a lot of us, and has been for a lot of surgeons for a long, long time, was that there was virtually no accent on tradecraft. It was all your brilliance, and your diagnostic ability, and knowing the right operation to do—but not knowing *how* to do it."

This became very clear to him when he worked for six months as a registrar, similar to a chief resident, in plastic surgery in New Zealand, under a gifted surgeon—"a man," Roger says, "who understood the mechanics and the biology of living tissue in children." Throughout his training, he says, "I was exposed to the full range, butcher surgeons—big talkers, big writers—and superb technical surgeons who understood all the materials they were using." As Roger spoke, I noticed how he underscored the words *the mechanics and the biology of living tissue,* and noted, too, his emphasis on understanding the materials—both physical concerns, as opposed to medical or clinical ones.

Attention to tradecraft, as he calls it, is situated at one end of the spectrum of this work, and at the other end is ethics, just as important as and yet the exact opposite of the physical facts of good or poor craftsmanship. To Roger's mind, he tells me, "one of the biggest ethical issues—it's been there from the beginning, it'll be there till the end—is informed consent." The patients he cares for are for the most part children and babies, who as a practical matter have no say in whether they're treated or how. Often the patient being discussed hasn't even been born yet. "The parents are the guardians of the child, the protectors of the child, and some protect better than others. They need to be informed, and they need to be able to provide informed consent for whatever's going on," he says. It's an ethical problem because congenital heart defects are so complex that it's all but impossible to convey what's involved in such a way that the parent will have a full understanding of what fixing the defect will mean. Because he can never know how much any parent is truly taking in, he tries to be as clear as he can. This is why he always says, "Risk *means:* death or near-death resulting in severe brain damage."

As a doctor, too, he's often cast in the role of counselor, with parents looking to him for the answer to a question that's typically phrased this way: "If it were your child, what would *you* do?" Roger believes in some ways that this should be the acid test for all doctors' and surgeons' decisions about what procedures to recommend. At one weekly catheterization conference, on a day when all the senior cardiologists were away at a convention, Roger asked Geoff Lane, Adel Younoszai, and John Rhodes (a third junior staff cardiologist), all of whom were under forty, what course they would advise taking in a difficult case. A six-year-old boy with congenitally corrected transposition of the great arteries had been brought to the clinic for diagnosis and possible treatment. In this rare lesion, the ventricles have inverted themselves in an effort to restore the proper circulation scheme, but the mutation doesn't work well in the long run. Attempts to repair the defect used to carry a high risk of surgical mortality and a grim long-term prognosis. Then, in mid-1989 Roger, practicing in Australia, and, shortly before him, a surgeon named Professor Imai in Japan independently performed the same operation, which has since come to be called the double switch, to fix congenitally corrected TGA; each was unaware of the other's work, and each had good results. Roger has now done forty-seven double switches with no mortality. This child's parents had come to Cleveland to get his advice. The difficulty was that the child was doing fine; by all appearances he was perfectly healthy. He would likely get worse, however (though maybe not), and by age twenty he might be failing altogether (or not). The thing was, the repair had never been terribly successful on older kids; it worked best on younger ones, with the odds for a successful result falling each year parents waited. So should a father check his healthy six-year-old son into the hospital to have major heart surgery when nothing was overtly wrong with him, and risk death or brain damage, bleeding, infection? Or should the parent wait to see if the child ever *did* get sick and then, if he did, hope it was still early enough that the surgeons would be able to save his life, preventing the slow, miserable heart failure that would otherwise kill him just as he reached adulthood?

Geoff Lane suggested that the switch should be done now. Roger turned to him in the darkened conference room, an angiogram of the boy's heart projected onto a movie screen, and said, "All right, Geoff, this is *your* six-year-old child." He listed the possible courses. "Are you still going to offer the double switch?"

Hedging somewhat, Geoff said, "I'd talk to you and get your input."

Roger said, "If we're going to do something, I'd want to do it now rather than four or five years from now. *And he'll do well for four or five years. . . .* I don't think we have enough numbers here to say. The older they are, the harder the operation, and the harder the operation, the more risk." He noted that so far, the data had been dependent on associated lesions. Then he added, "It's very hard to take a well kid and tell him we're going to do a big procedure."

Roger described his bottom line to his young cardiologists: "If we're not going to do it on our own kids, I don't think we can recommend it to parents. Our acid test must always be, would we do it for our own kids?"

This led to a discussion about counseling: How should a cardiologist review the options for parents who knew little about complex heart defects or the short- and long-term implications of surgery? "When you offer something, it depends how you offer," Roger told them. "We're a nation of salespeople, and you can sell what you like." He had said much the same thing to me earlier, mimicking two doctors, one who wanted to get a parent to take a chance on a risky surgery, and one who wanted to get the parent to decline the risk: " 'There's a fifty percent chance we can get your kid through.' " Roger nodded and spoke these words with an upbeat tone and a smile. Then his lips curled inward, his brow furrowed, and he said, " 'There's a fifty percent chance your child will die.' " Reverting to himself, he stated the obvious: "It's the same information."

Advocating for a high-risk treatment using Roger's acid test should not be confused, however, with responding to a parent's request for *moral* direction, he said. The distinction became crucial when compassionate care was at issue—the policy of doing nothing for a critically ill child, allowing him or her to die as comfortably as possible. "That's an easy one when we have no surgical options to offer," Roger noted. "You just say, 'There's nothing we can do.' " The situation became more complex when the surgeon *could* do something, but only at a huge risk; in such cases, he said, the parents must be clear about the risk involved. "But if that risk evolves from a ninety percent risk into a ten percent risk, should the parents still have the right to say, 'I don't want anything done'? And where's the breakoff point? I don't know the answer to that."

Roger told me the story of a family who had a baby with hypoplastic left heart syndrome (HLHS). This was at a time when the Norwood operation was still young, and Roger was quoting a

50 percent mortality rate. So he presented two options: Norwood or transplant. In this case a heart became available. He went down to the sleeping room in the parents' lounge, woke up the mom and dad, and explained the situation to them, then told them he needed to have their decision within three hours. They were counseled about the benefits and drawbacks of each, and the long-term implications. In the end they opted for transplantation. The operation was successful, but over the year that followed, as he learned how difficult it was to care for a baby with a heart transplant, the father became furious. He threatened to sue Roger. A few years later the father brought in the child, now gravely sick from transplant rejection, too far gone to be helped. The child died—a death Roger suspected the parents could have prevented had they brought the child in sooner.

Compassionate care for babies born with HLHS was routinely offered as recently as a few years ago, and it may still be offered today at smaller centers. The Norwood operation has allowed many such babies to live, but mortality and morbidity are common, and recovery can be excruciating. At first many centers didn't believe in presenting this high-risk surgery as an option; now some surgeons feel it's unethical to offer compassionate care. At one recent conference, for example, Richard Jonas, the chief surgeon at Children's Hospital in Boston—an institution considered by many to be the standard-bearer in the pediatric field—argued against its provision. I asked Roger if he favored preserving compassionate care as a possible choice for HLHS babies.

"Yes, I do," he said. "If you've got a family that's really stretched economically, not very smart, and they've got five kids, what are you doing if you don't offer compassionate care or don't give them that out? You're forcing them to add a sixth child who's going to totally destroy any reserve they've got and affect all the other kids. I mean, good gracious.

"So these families are going to struggle along, and if they don't recognize that their kid's gotten sick, and they probably won't, and they don't bring him in until the kid's moribund and he dies, then you've also imposed that death on them. Whereas there was an opportunity to feel that death was a very reasonable alternative.

"I'm afraid I think one of the huge burdens periodically of being a parent is making these decisions, and there's no easy way out."

Moreover, *his* moral vantage was no more valid than that of the parents, Roger said: "They've got to dig deep, they've got to use their own support structure—the grandmother, the grandfather, friends, if they've got them—and talk to *those* people about the human issue. The decision has to be theirs. But what I do tell them is that whatever decision they make, we'll give them our support."

It was Roger's job to make sure the parents were as fully informed as possible about the medical issues and to give them a set of clear options as to what they might do; it was his job to recommend intervention only if it was the course he himself would choose for his own children; and it was his job to carry out, to the best of his ability, the parents' wishes.

Ultimately it must be their choice. But how could anyone know the long-term implications of such a choice? In the case of Drew H., who had struggled through four months of life, whose chest had been opened, whose heart had been cut and stitched, whose vessels had been filled with every manner of necessary hardware, the parents had asked Roger to open their baby boy's chest again and see if he could do something, anything, to keep him alive. They'd made their decision in the beginning and again at every step along the way. Roger told them he believed he could deliver blood to the left side of Drew's heart, though he didn't know if the heart would respond. It was a complicated case, but he was hopeful. At this late stage in the baby's illness, transplant wasn't a realistic option, and so it was never discussed. Roger would try for surgical repair.

"Bad start, really bad start," Roger mutters again as he investigates the heart, looking for the coronaries, which are obscured now by all the scar tissue. Fackelmann, his regular assistant, scrutinizes the heart, too, using his sucker to pick up the continuously oozing blood. Roger says nothing more as he locates the coronaries and then begins to take down the original switch. With the cross clamp on, it's especially important not to waste time. He can't know how well a heart this sick will tolerate being stopped for the time it takes to redo a switch.

He begins by transecting the main pulmonary artery and the aorta at their respective suture lines. He sees in the aorta the homograft hood and removes it to reveal the two coronary ostia. The right ostium is clearly open, but on the left there's something that looks more like a dimple where the opening ought to be—it's blocked. Roger asks Bob Cherpak for a one-millimeter probe, which he then manages to push first through the left coronary ostium into the artery itself. By cutting along the inner wall of the aorta, where the intramural section of the coronary runs, he makes a one-and-a-half-millimeter opening for the left coronary. He then cuts the coronaries out of the aorta completely (allowing each a generous cuff), cuts new flaps in what used to be the main pulmonary artery, and begins to sew the coronaries in again.

Roger rarely speaks during this procedure. Once, when Charlie, to his immediate left, fails to understand what he needs, he says, "That sucker is pointing the wrong way, get it the hell out of there," and Charlie removes the sucker and watches. Another time he tells him, "You're not looking, you're costing us valuable seconds." When Kevin, manning the perfusion pump, reports, "That's twenty minutes of pleging," Roger says, "Oh, Jesus," the words barely audible behind his mask. The effects of cardioplegia are temporary, so every twenty minutes Kevin must run more plege, which Roger must direct into the coronaries. Each time Kevin makes his announcement, he's also reminding Roger how much time has passed, how much less he's accomplished in the last twenty minutes than he'd hoped, how much more sewing remains to be done. "Jesus," he whispers again, shaking his head. And Kevin repeats, "That's twenty minutes of pleging." And repeats it once more.

Drew's heart has been cross-clamped—stilled, a flaccid, bloodless muscle, empty, nearly white—for more than an hour now. The O.R. is quiet save for the rush of air-conditioning that keeps the room cool (about 60 degrees), the hum of the bypass machine, and the door whooshing open occasionally as a technician or nurse comes or goes. There is no heart rate to monitor, no breath to give to the patient. From the windows of the O.R., cars arriving at and departing from the front entrance are visible in the crepuscular light; their headlights illuminate the driving snowflakes and the deep grooves of tire tracks in the snow. The snowfall has turned into a winter storm.

"Start warming," Roger orders Kevin. He has almost finished reanastamosing the aorta and is therefore nearly ready to take the cross clamp off, let blood fill the coronaries, and hope the heart starts; he'll complete the operation with the heart empty but contracting. But when he removes the cross clamp, allowing blood to flow to the muscle itself, the heart does not begin to beat. Instead, it fibrillates. Roger asks Julie to administer amiodarone, a cardiac depressant that suppresses arrhythmias. "Table away from me," he says, and Julie presses a remote control that tilts the table. "More, more, more," he says. "I want a sustained blow." Julie gives the bellows a squeeze, filling Drew's lungs with air, and holds it. "Half flow, please," says Roger. He rubs the left ventricle, massages it, then slips his right hand under the heart. It's unresponsive to the drug, so he asks for a shock. Charlie gives him the paddles, and he delivers the *pop*. The heart continues to wiggle, contract only sporadically, or move not at all. He gives it another shock; nothing. And another shock.

"What's the perfusion pressure?"

"Forty-five."

Drew's heart is all but motionless. "Deb?" Roger calls, wanting to shock it again, and Deb hustles back to the generator behind him. He places the paddles, but before he asks for the jolt, the heart begins to beat on its own.

Roger can now go on with the procedure, patching the back of the aorta with a strip of the baby's pericardium and then sewing this part of the aorta to the main pulmonary artery.

Debbie has paged Dan Murphy, who now arrives wearing a gown and a pullover cap, fastening the bottom tie of his mask as he enters the O.R. He rolls the heavy echo machine close to the head of the table, between the anesthesia and perfusion machinery, pulls up a stool, and plugs in the echo probe. He leans close to look for a good view, his back to the patient, his right arm reaching back to control the probe as he watches the black-and-white static on the screen. "The heart has no blood in it," he says, "so it's hard to tell." He'll wait till they send blood back into the heart before making any guesses about the success of the operation.

Dan stands up and peers over the cage at the baby's open chest. "It's pretty empty, Roger," he says. "It's wiggling, but it's not ejecting."

Roger asks, "Did you notice any AI?" He noted after he took the cross clamp off that the heart distended, suggesting that the aortic valve wasn't working properly—a condition called aortic incompetence.

"A little," Dan replies, "but not enough to affect pleging."

Roger tells him, "I had to start from square one, redo the switch. As for the intramural, we found it. It was blocked."

When Roger has finished his sewing, and the heart fills with blood, Dan sits back down on the stool to get more pictures of the organ, this time focusing the probe on the left ventricle. Roger cranes to look over the cage at the echo screen and the slow undulation of the ventricle.

"Is that real time?" he asks, hoping Dan has put the image in slow-motion mode.

But Dan answers, "That's real time."

"That looks pretty awful."

"Yeah, and it's pretty global."

"Did you see any flow in the left coronary artery?"

"We really didn't see any before, but I'll take a look," Dan says. Then, "Roger, I really can't see flow in the left coronary artery. I *can* see flow in the right."

Roger stops working when he hears this. He tilts his head back and then to the right to stretch his neck. He closes his eyes, takes a deep breath. After a few motionless seconds he opens his eyes again and stares at the heart. He looks up at the monitor suspended from the ceiling off to his right. He looks back at the heart. He says, "If the left coronary's not working, we're fucked."

Roger began this day as he usually does, with eight A.M. rounds. These were followed by his first surgery of the day, a Norwood on a newborn, still one of the highest-risk procedures performed. This case was especially difficult because the baby girl was not a typical hypoplast but instead had a hypoplastic aortic arch, aortic stenosis, and heterotaxy (an abnormal arrangement of the heart structures), and on top of all that was in heart block (meaning that her heart was malfunctioning electrically). Drew was his second case. It's now dark outside, and Roger realizes he's got to do yet another time-consuming procedure on this baby: the takedown of a mammary artery and a bypass graft onto the left coronary artery. This is when it gets hard, he says—when your first attempt doesn't work, and you have to go back in and start over. Wasting no more time, he begins preparing to graft the internal mammary artery onto the left anterior descending.

June Graney descends to the third floor, passes a large central aquarium, turns the corner, and heads to the parents' lounge to give Angie, Bart, and the rest of their family the news: Dr. Mee doesn't know yet whether or not he's been able to fix the intramural coronary; the ventricle still isn't working well, so he has to assume there's no flow, and he's going ahead with a bypass graft, using a mammary artery.

Although this is an emotional setback for the family, they've prepared themselves for the possibility and therefore aren't devastated. Angie is more upset than Bart, who tends to be pessimistic by nature, always expecting the worst.

An observer would scarcely be able to tell that anything was wrong, though, from Bart's appearance. His expression is blank, and he's not much of a talker, anyway—certainly not when it comes to sharing his feelings. Bart's a *guy;* he works outside building fences in farm country for a living; he's not openly reflective, doesn't show much emotion; he loves auto racing and hunting, has been a hunter since he was ten years old. As Angie puts it, "Guys have a different way of dealing with things." She doesn't mind his quiet nature and lets him go his own way when he needs to.

A few days before they were to leave for Cleveland in this last-ditch attempt to save their son, Bart asked Angie if it would be OK for him to go hunting. The rut was picking up—the time when the does are in heat and the big bucks come out—and it would be his last opportunity of the season for hunting deer. Angie encouraged him. The past four months had been just as hard on him as they'd been on everyone else—maybe even harder, since he had no outlet for the fear and frustration of watching his baby boy grow sicker and sicker by the day, and knowing they'd soon have to head to another city so he could have more open-heart surgery. Angie knew that for her young husband, the best part about hunting was simply being alone in the woods, where he could think and sort things out for himself. Before he left the house he leaned over his son and said, "I'm gonna bring you home a monster buck." Then gave Drew a kiss and departed into the darkness.

He always said that to Drew, and he always meant it. But the truth was, he'd never gotten a big buck before, had never even seen one in all his fourteen years of hunting—at least not closer than three hundred yards, well beyond the range of his Hoyt compound bow and the 90-grain Muzzys he liked to use. He often complained to Angie that it seemed like he wouldn't ever get that big buck.

This early morning he drove and then hiked to his usual tree in the woods. He wouldn't have that much time here before he had to leave for work. "Although I had only a few hours to hunt," Bart recalls, "a few hours during the peak of the rut is better than a lot of other things I can think of." Soon after he removed his grunt call at daybreak, a small button buck appeared less than twenty yards away. Bart watched him from his stand, the buck lingering till Bart had to leave. He'd be back one last time tonight after work, before the season ended for him and he and Angie took up temporary residence in an ICU, so they could watch over their sick son.

Bart was back in his stand at 3:35 that afternoon. It was a quiet November day with a light breeze. He kept still for twenty minutes, "allowing the woods to calm after my walk in," he explains, and then issued a series of grunts. He listened to the squirrels darting about in the dry leaves, stared at the laminated photograph of Drew that he'd affixed to his bow. And then he heard the sounds of a deer behind his setup. The hoofbeats came closer, and then he heard it: three grunts. He didn't need to see it to know it was a buck, a big one. Slowly, he drew.

The deer came into view, running, on Bart's left. He *was* big—a monster buck. He stopped forty feet away and lowered his head to explore some of the scent Bart had put down that morning. Bart was at full draw, but as yet he'd had no opportunity to shoot. The buck was facing directly away from him—still no shot. If the animal bolted straight ahead, Bart would lose him. He had to take the risk: he grunted, and the buck turned enough for him to get off a quartering away shot. The second he released the arrow, the buck ran. Bart knew immediately that it'd been a perfect shot, a clean, fatal hit, and the copious trail of blood left by the buck confirmed it. As the buck wandered in a wide circle around his stand, Bart left the woods, so as not to spook him as he died. He called out to his uncle, who'd joined him today. He was quiet, but inside he felt elated and amazed, his heart beating powerfully as he and his uncle returned to his setup, then followed the blood to where the deer lay. It was a monster buck and more—a once-in-a-lifetime, record-book buck.

Bart began to weep. The buck came for Drew. He doesn't usually show his emotions, Bart admits, but he wept openly that day, when he beheld the awesome animal. "I don't believe it was luck," he says. "I believe the buck was dead before it entered those woods—spiritually." As a boy he used to steep himself in tales of Native Americans and their myths, and he will tell you that the Indians believed the white-tailed deer was the spirit of hunting, that they drank its blood to become better hunters. And this was his conception of that deer, too—for Bart, the deer was a mystical creature.

He waits now in a parents' lounge in an imposingly vast hospital in an unfamiliar city, far from the woods, and his son lies immediately above him in the O.R., the first part of his operation over and unsuccessful. Drew's relatives remain hopeful and hunker down in front of the television, or with their cards, and redouble their prayers. They stare out the window at the snow that's really coming down now, at the beams of headlights moving through the blizzard below them.

· · ·

When the heart is beating steadily, an anastomosis of the mammary artery is in itself neither a risky nor a complicated procedure; it merely requires painstaking care in the exact artery-to-artery stitching of two vessels that are each less than a millimeter in diameter, in such a way that when the stitching is done, the juncture will remain open, allowing blood to course freely from the mammary into the coronary. Roger has already found the artery, ensured that blood is flowing through it, taken it down (freed it from the chest wall), and closed off its branches with tiny clips. He spends about half an hour preparing the arteries for the work he will do. Because he wants to avoid stopping this sick heart for any longer than is absolutely necessary, he intends to do the sewing as it beats. He gently tugs the left anterior descending (LAD), the critical coronary, off the heart and suspends it above the beating muscle with ligatures to keep it more or less still. He then begins carefully sewing the mammary at an oblique angle into the proximal LAD. Blood will flow through this artery directly into the LAD, which feeds the heart.

As Roger's sewing, the large head of Jonathan Drummond-Webb appears over the cage. Having finished his own case hours ago, Jonathan has slipped into the O.R. through a side door. He rarely ventures into his partner's O.R. anymore. Occasionally he'll assist, or he may look in just out of curiosity, but tonight he seems unusually quiet. He doesn't stay long and hardly speaks when he's here, perhaps because of the gravity of the case. He asks Roger how it's going.

Roger responds without looking up, "Well, I redid the switch, but it doesn't look like it's working."

Jonathan stays only a moment or two longer before disappearing.

The work is slow, and it's complicated by persistent bleeding; because so many vessels have grown from Drew's pericardium into the myocardium, the two layers they've had to separate, there's a good deal of bleeding that's taken extra time to control. It's all but impossible for an observer to see what Roger is doing at this point, he's working with such tiny structures. Since he's got at least another hour's worth of work ahead of him, I leave the O.R. to see what's happening elsewhere.

The corridor, a long strip of blue carpeting flanked by rows of secretaries' desks, behind which are the doors to the cardiologists' offices, is already deserted, though it's not yet six o'clock. Anyone who didn't need to be here has bolted because of the storm, and the lack of activity makes the hour feel later than it really is. *Nothing* seems normal; there's a bad-weather feeling in the air in here because of the unusual dark and quiet. And there's actual bad news in the ICU: after seven weeks of waiting for a heart, Ashley Hohman has taken a turn for the worse. Over the past few hours she's gone from needing little maintenance, relatively few drugs, and steady but controlled ventilation to the opposite: she now requires lots of drugs for blood pressure and cardiac function, as well as increased ventilation assistance. These are the first signs of a sepsis tailspin, something everyone here has seen before—too recently, in fact. Just three days ago Julia Stone became septic and died within twenty-four hours. There's nothing Marc Harrison, the attending on call tonight, can do for Ashley, other than keep driving the meds into her little body. Kelly is by her side, and Tim, who got the news by phone, is even now hurtling up I-71 on his way here. Ashley's belly is extremely swollen from the acites, and her skin has a pale purplish cast.

Even if a heart became available tonight, it wouldn't be in time to save her. And they wouldn't be able to get it here anyway because of the storm, Marc says.

"No one should be sent out in this weather," he tells me. He personally knows of two transport missions that turned deadly. When he worked in Bangor, Maine, his hospital shared a transport service with a hospital in Portland. One day the transport helicopter took off from Portland to pick up a burn victim in an outlying town and got caught in bad weather; the pilot overestimated his abilities, perhaps, and ran out of gas over the Atlantic, where the helicopter went down. All aboard died, according to Marc, except for the pilot, who somehow stayed alive in the freezing seas by clinging to the wreckage. And when Marc was in Salt Lake City, a helicopter from the University of Utah Medical Center was dispatched to retrieve someone who'd been injured in an avalanche in a canyon in the Wasatch Range, whose highest peak exceeds twelve thousand feet. On its return trip, the helicopter ran into a wall of the canyon, killing everyone on board, including the patient, who'd been in serious but not critical condition. Marc had known the whole team, but the flight nurse had been a particularly admired friend.

Such stories are not uncommon. Paula Bokesch, the anesthesiologist who began Drew H.'s case five hours ago, was working at Children's Hospital in Boston when a transport team, responding to a call from an organ-procurement organization, flew into a winter storm to harvest a heart on Cape Cod. This plane went down, too, killing the whole team—and wasting the heart, she notes.

Unable to leave the hospital because of the snow, Paula is reminiscing in the O.R. lounge. Abud Neto, a surgeon from Brazil who's here to observe Roger for a month, is in the lounge as well. His English is not good, but he understands enough to nod and add that he, too, lost a colleague in a transport accident. (Among the many tens of thousands of air medical transports this year, there will be twelve accidents, five of them fatal, according to Rick Frazer, who studies and writes about air medical safety and is operations manager for Lifenet–Rocky Mountain Helicopters in South Carolina.)

I continue to wander as the evening stretches out in this eerie way, lending the place a kind of no-one's-going-anywhere insularity. Kathy Weise, who was today's attending in the ICU, returns to the unit carrying a Pepsi and a Pizza Hut box from the food court downstairs, which is now mobbed by stranded people. A couple of hours ago Kathy got in her car to go home, pulled out of the parking garage, and turned left onto Carnegie, at that hour typically a busy but quickly moving six-lane, one-way thoroughfare. Today it was all she could do to nose out into the nearest clogged lane—and she regretted it almost immediately. She spent an hour listening to National Public Radio, saw that she'd traveled only two blocks, took a left, and started back down Euclid, on which she was able to enjoy another hour of National Public Radio before reaching the parking garage.

In the O.R., as Roger Mee ties the final stitches of this bypass graft, an attempt to get blood to Drew's left ventricle, Dan Murphy is once again paged to reecho.

Dan strides in, sits on the stool in front of the echo machine, and reaches back to grip the echo probe. No change. None whatsoever.

Dan stands and faces Roger to tell him the news: the function is still terrible. Roger is at this instant busily trying to control all the bleeding from the vascular adhesions.

"Could you see any flow from the coronary?" he asks.

Dan tilts his head and says, "I didn't see any flow, but I'll try again." He knows he won't find any, but the results of that last echo were so utterly discouraging that he's willing to give it one more try.

At last Roger pauses. He's been moving pretty much continuously for the entire five hours he's been hunched over this baby, but now he stops, straightens his back. He looks at the monitor, glances around the table, then looks down at the open chest before him. He sags. He abruptly sags: his shoulders contract, and his upper body drops an inch. It's the moment he admits defeat.

Then he lifts his head and says clearly, "I think we've failed with what we were trying to do."

These words are jarring. I've never heard him sound so discouraged, so beaten. What he's saying is that the future of this child is no longer in his hands, and Drew's prospects are actually worse now than they were before he was brought into the O.R. He's saying, *I've* failed. He's particularly hard on himself, tending to think the worst. The fact is, there may be some blood flow to the left ventricle; he just can't be sure, and it will take days for the heart muscle to prove itself viable. If it doesn't, of course, the outlook for the baby is grim.

Roger knows he can't get this kid off pump without giving him massive doses of inotropes, so he decides to put him on LVAD, a left ventricular assist device. This may be the end for Drew. The team begins preparing for this final stage of the procedure, June Graney is debriefed so she can convey the news to the family, and Roger shakes his head, frustrated. "I should have known better than to spend all that time on bypass trying to put a graft in," he says.

After Roger utters those words, a few minutes pass and the O.R. is quiet. Then Jonathan strides into the outer vestibule and punches the metal plate on the wall, and the O.R. door opens with a *whoosh*. *"Roger,"* he says, *"I may have a heart for your boy there."*

Roger is bent over the table working, and he doesn't turn his head to look at Jonathan but maintains his focus on the field, still trying to control the bleeding. From where they're standing, Frank and Mike Fackelmann can see Jonathan without having to turn.

"Oh, really?" Roger says with what sounds like British understatement. Then he asks, "What about Ashley Hohman?"

"I'm not doing her now, because she's septic."

"Why?"

"She's *septic.*"

Mike Fackelmann can't believe what he's hearing. *"No one's going to get a heart right now,"* he says.

"I'm not talking to you!" Jonathan shouts at Fackelmann. *"Roger?"*

The last transport run Mike and Frank went on, just two weeks ago, included a hairy landing in a lightning storm that had even Frank clutching the armrests, second-guessing his love of such missions, and hoping he'd be around to forget this one. And then last week the adult transport team had its own near-disaster when its plane depressurized and had to make an emergency landing. Mike's buddy and counterpart in adult cardiac, Larry Forbes, told him about it by way of explaining how he'd smashed his face and broken a tooth. So "No one's going to get a heart right now" seems to me a not-unreasonable comment, given the conditions outside.

While Roger is doing a watchmaker's dispiriting business in the baby's chest, he must make a decision: should they wait to see if Drew's heart muscle will come back, not subject him to transplant, but risk his dying soon if the procedure hasn't worked, or should they go immediately for transplant, pursue the heart, and deny it to some other on-the-brink-of-death infant. Roger's mumbled answer is caught in his mask as he continues to work without looking up.

Jonathan cups his hand to his ear and leans all the way over in a caricature gesture, a grimace on his face. Frank looks up at him, nods vigorously, and winks; I can all but see a giant grin on his face. Roger has said OK.

"I'm going to go for it," Jonathan says, and he heads for his office.

Going to go for it. It's not a done deal. Maryanne Kichuk, the head of the transplant program, has "heard there's a heart out there tonight," Jonathan explains to me.

The way he phrases it makes it sound like a rumor. What does he mean, Maryanne has "heard" there's a heart "out there"? Who did she *hear* it from? Out *where*? Is it just sort of floating out there somewhere, a little winged heart fluttering in the clouds?

June Graney and Corinne Conner, the two cardiac nurses on tonight, both in their customary blue scrub pants and blue scrub jackets, follow Jonathan to his office, hoping to learn what's going on. But Jonathan's not saying anything until he knows more. June asks how he expects to get a transport unit out of the clinic; with all the traffic at a standstill and the snow relentlessly piling up, how could it get to the airport, let alone fly out in this weather?

The question annoys Jonathan. He's hopped up and tense. Maryanne is no longer reachable, apparently, so it's up to him. "We'll get a helicopter," he snaps. "I don't care how we do it. That's not my job." He stares at the computer on his desk, one hand on the mouse. When June and Corinne leave, he ostensibly begins to "look" for the heart that's "out there tonight."

Maryanne Kichuk was stalled in traffic when the call came in on her dying cell phone. The distance between the clinic and her home on Cleveland's eastern edge was only a couple of miles, but it was an *uphill* couple of miles toward the Heights, and nothing was moving. Maryanne, a forty-year-old cardiologist with angular features and straight, dark hair, had been enthralled by pediatric heart transplantation ever since she was a second-year cardiology fellow at Columbia Presbyterian Medical Center in New York City, where some of the first pediatric heart transplants were done. Since hiring her three years earlier specifically to develop its transplant program, the clinic had transplanted about thirty patients, with just one surgical death; at an average of about ten to twelve pediatric transplants a year, hers was a medium-size program with good numbers. This was what she'd always loved, but she'd never experienced a night like this. Neither, for that matter, had Phil Kozell. It was his first night on donor call, and a call actually came in from LifeBanc, the organ-procurement organization serving Cleveland. LifeBanc had been contacted by UNOS, the United Network for Organ Sharing, a nonprofit, federally licensed agency that monitors organ procurement nationally. A viable heart was available in the southern part of the United States, and UNOS had determined that it was a match for a patient on LifeBanc's list. The patient was Ashley Hohman at the Cleveland Clinic. LifeBanc gave him the details, and Phil—his first night on!—phoned Maryanne.

Straight off, Phil said, I don't know, Maryanne, is this heart too big for Ashley? And what kind of condition is she in?

Maryanne admitted that her situation wasn't good. Jonathan had been tapping Ashley's belly because it looked as if she were going to explode. She was severely hypotensive—a likely indication of sepsis—blue as a blueberry, and probably close to dying, maybe before they could even get the heart up here, let alone get her into surgery. But Maryanne nevertheless told Phil she'd try to reach Jonathan to check with him. She was inching along Fairhill, one car in a mass of them creeping up the long, winding hill, some of the vehicles already abandoned or unable to proceed. She got Jonathan on the phone and informed him that there might be a heart for Ashley—did he want it?

Ashley wasn't in any kind of shape for anything, Jonathan answered. And it sounded as if the heart would be too big for her anyway—I won't be able to fit it into her chest, he said.

Then there was a pause, as if the same idea were occurring to each of them simultaneously. *Drew H.*, they thought. The baby who'd presented so urgently yesterday, his left ventricle ejecting only 8 percent, practically dead, and who was at that moment still in the O.R. Jonathan said, "Let me see how Roger's doing in there. I'll call you back." He hustled to the O.R., put on a hat and mask, slipped in through a side door, and poked his head up over the cage. In answer to his question, Roger told him, "Well, I redid the switch, but it doesn't look like it's working." Jonathan stayed a moment longer, read the blood gases, watched the monitor, and knew what Roger also suspected: this little boy would never make it off pump. He returned to his office and called Maryanne. That baby isn't coming off anything, he said. Can we get the heart?

The blood type and size were a match, but neither Maryanne nor Jonathan knew how this plan could possibly work. Drew H. wasn't even *listed* for transplant. Organs for babies were such an incredibly valuable commodity, protocol for getting them was necessarily rigid. She told Jonathan that first they had to get the baby listed—

Then her cell phone died, and she was alone in her dark little capsule in a snowstorm, hoping she could make it up the hill. It would be hours before she learned what happened next.

It was at this point, after Maryanne's cell phone died, that Jonathan returned to the O.R. to tell Roger that he might be able to get a heart for the patient on the table—should he go for it? After getting Roger's OK, he went back to his office, trailed by June and Corinne.

First, he thinks, he'll need to call Phil and tell him they want the heart, not for Ashley, but instead for another baby who isn't on the transplant list. When Jonathan lays it out for him, Phil says he has no idea if that's possible. He calls his boss, Renee Bennett, a former cardiac nurse and now the head of the organ donor program at the clinic, who's just made it home. He describes the situation for her and asks what to do. We can do it, she says, but we have to get him listed, first. Then she phones the Georgia organ-procurement organization (OPO) to get things moving.

June in the meantime—*This is crazy,* she says to herself, *it's happening too fast, what is going on tonight?*— is thinking about Drew's parents. They haven't been told about the donor heart yet; in fact, no one's even *talked* to them about a transplant. Deciding to consent to such a procedure—making an *informed* decision, giving *informed* consent—takes time. A hospital can't just pop in a new heart and send the family on its way. Caring for a baby with a transplanted heart isn't easy, not for the baby and not for the family. The child is still, essentially, chronically ill. As one doctor put it, "Transplant is exchanging one heart disease for another." No one's required by law to accept a donor heart—parents don't *have* to agree to a transplant for their baby if the doctors suddenly announce that a heart is available. Although it's now performed routinely, the surgery is nonetheless still considered experimental. The family has to want it, and that decision must be based on clear information. But there's been no discussion with Drew's parents about what transplantation would mean. The only reference they've heard to it was when Dr. Murray, weeks ago, mentioned it, as something that Roger might (but didn't) recommend. Angie and Bart, pacing in the parents' lounge, have *thought* about transplantation, talked about it with each other, but never in these terms; rather, they've resolved that if they lose their son, they want his organs to go to a baby in need.

June descends to the parents' lounge to talk to Bart and Angie. She doesn't mention the heart, because it's still unclear whether they can have it and whether transport will even be possible. Instead she carefully explains that the operation has not been successful so far, and that Dr. Mee is preparing to put Drew on a left ventricular assist device. Given this, she suggests that they should try to get him listed for heart transplantation as soon as possible— tonight—rather than waiting any longer. Is this something they'd like her to do? she asks. Later, she can talk with them some more about what all this means, but right now it would be a good idea to begin the listing process. Angie and Bart give her the information she'll need to relay to Phil so he can get Drew onto the UNOS list and then officially accept the heart that's "out there." June says she'll ask Jonathan to come down and give them a better sense of what transplantation might mean for the family in the long run.

When Jonathan speaks with Renee, he learns that she has already been in touch with the Georgia OPO. It turns out that Ashley is not the first patient on the list but rather the *backup*. Each time a heart comes up, UNOS calls the OPO of the first patient on the list who's a match for it, and then, in the event that any number of possible complications should arise, the organization notifies a backup, so that two teams will be ready and no time will be wasted in harvesting this most precious commodity. In this instance, Renee says, the doctors for the first baby on the list intend to take the heart, which is now destined for a center in Georgia run by a surgeon I'll call Dr. Smith.

Jonathan is committed to getting this heart. He calls Smith, reaching him at the hospital. He describes the situation to him, runs through the details of the case. Smith says he doesn't like the sound of it. He tells Jonathan he's concerned that the patient's history, with all that time he spent on ECMO, makes him a poor candidate for transplant. (Don't feed me that line, Jonathan thinks.) His own patient, Smith relates, is a little hypoplast baby who had a failed Norwood and has been waiting weeks for a heart. The patient is already being prepped for surgery. Jonathan counters that Drew H.'s condition is critical; he'll likely die if he doesn't get this very rare heart. Is your patient stable? Jonathan asks Smith. Will your baby survive without this *particular* heart? Smith agrees to talk to the cardiologist in charge of the case.

Together Renee and Phil are meanwhile working on listing Drew.

Smith calls Jonathan back. You can have the heart, he tells him. Jonathan thanks him.

Now the question becomes, can they get a team out in this blizzard? Some of the highways are closed—are flights even leaving Cleveland's airports? This is Phil's job. He puts in a call to Eagles Wings, the flight company that's contracted with the clinic to lease it jets for organ harvests.

In the O.R., while Jonathan is on the phone with Maryanne and Phil and Dr. Smith, Roger is trying to achieve hemodynamic stability, meaning that Drew is still bleeding like crazy and they've got to get him on an LVAD—put a right-angle tube through one of the pulmonary veins and into the left atrium, where it will suck up the blood returning to the heart from the lungs and pump it out through another tube into the aorta, thus bypassing the left ventricle. No more than ten minutes have gone by since Jonathan delivered the news of a possible heart when Drew's heart fails. It doesn't even fibrillate; it just slows down, then stops.

Roger grabs the paddles, calling, "*Deb*." Debbie Seim hurries over to the machine, which is sitting behind Roger, on the lower shelf of a rolling cart. She charges it to a level of 5 joules. *Pop!* Nothing; the heart has stopped.

"We're in trouble," Roger says. He places the paddles on the heart again, and Debbie cranks the knob to 10 joules. "Shock again." One more *pop,* and then the right ventricle picks up its steady rhythm. Roger stares at it. It continues to beat. He says, "I don't know what happened there."

He goes back to work on this heart that has been open on the table for more than six hours now. Fackelmann is noticeably reserved. Roger now begins to hook up the LVAD, even as he tries to keep the bleeding under control. He says, "I think the world is coming to an end. The U.S. doesn't have a president, and there's a bloody bad snowstorm." By now even the crew in the O.R. is buzzing about the weather, wondering if we'll all be stranded in here, commenting on how the McDonald's is backed up with people terrified of running out of food. And a heart is on its way for the baby on the table. "I think Bush should just give it to Gore and be a gentleman about it," Roger concludes.

The mood in the room has gone from extreme tension to jokey ease as Roger secures lines in the patient. But forty-eight-year-old Debbie, the head O.R. nurse and tonight's circulating nurse, a fifteen-year veteran of the clinic, says, "This is the weirdest night I've ever experienced."

Ten minutes later Roger is finished. Before he steps away from the table, he stares into the chest in which he's been working for half the day and whispers, "I should have known better." He steps down from his stool, and Debbie unlocks his headlight cable from the light source. He removes his gloves and gown, signs the sheet attesting to the surgery he's just performed, and leaves the O.R. Once outside, he returns the headlight to the cabinet, stores away his surgeon's loupes in their padded wooden box. He washes his hands, then heads into the O.R. lounge and collapses into a chair.

June and Corinne arrive, having heard he's out of the O.R. He's been in there so long they've accumulated a list of issues they need to address. But both are quiet when they see him and wait for him to speak first.

"We didn't do well in there," he says. His face is pale but bright pink on the forehead and around the eyes and over the bridge of the nose from the pressure of the headlight, glasses, and mask. "It looked good from where we were, but Dan didn't see any flow."

June nods sympathetically. She pauses before saying, "Miss Hohman looks like she's on her way out."

"What is it, septic?" he asks.

June nods.

"Jeez," Roger says, tilting his head back. "The ICU is really taking a pounding."

"Yeah," June agrees, and waits another moment before addressing the other matters: beds available, who's being moved to the floor, who's being moved to adult, what cases should remain scheduled for tomorrow. And soon they're speaking of things that have nothing to do with work. It's past nine o'clock, and Roger's daughter is flying in from Boston tonight, home from college for Thanksgiving, the day after tomorrow. Roger says Helen was planning to pick her up, but given the weather and the fact that he's halfway to the airport already, he'll do it instead, if her plane can make it in.

After he's had some time to rest, he goes downstairs to see the parents.

The whole family braces when Roger enters the room. He sits on one couch, and Angie and Bart sit on the one adjacent, surrounded by their relatives. He's straight-faced, and so are they, already aware that the operation did not go as hoped and that they may now have no other choice than to list Drew for transplant, though they've never dreamed it would come to this. Roger explains what happened during the surgery and what the baby's heart function is. A transplant is now his best chance, he says. On hearing the word *transplant* from Dr. Mee, Angie begins to cry.

Then Roger gives her and Bart the news: a suitable heart has become available down south. Now, he says, what I need to know from you is, should we go get that heart?

He tells Angie and Bart they have two options: they can leave Drew on the LVAD for a couple of days and see if his heart function comes back, or they can move straight to transplantation.

Angie and Bart, who have been at the hospital for not much more than twenty-four hours, are now faced with a choice that would have astonished them only yesterday. And they must make this choice immediately. There's little time for information or consent. It's all happening so fast.

Angie says, Dr. Mee, what do you think? You was in there. You saw his heart. Do you think the function will come back?

No, Roger replies. I don't think it will.

He pauses, then adds, And there's a heart. Hearts this size don't come up that quickly.

Thus, right there, just like that, Angie and Bart decide: Yes. Please get the heart.

Before Roger says good-bye, Angie asks, Dr. Mee, will you be doing the transplant?

Roger answers, Dr. Drummond-Webb normally does the transplants. But I can be there to assist if you'd like.

Angie nods: yes, please. Then she says, Thank you, Dr. Mee.

Roger stops in his office to check on his daughter's flight. The airport is open, having been closed for only half an hour. But arrivals are badly delayed, in part because of the weather and in part because a number of flights have been redirected from Buffalo and Pittsburgh, whose airports remain closed.

Phil Kozell, in transplant control, has contacted the flight company, which has a pilot willing to fly if the team can get to Burke Lakefront Airport, a small field downtown on the lake, next to the Rock and Roll Hall of Fame.

After finishing up in the O.R., Mike Fackelmann and Frank Moga wheel Drew, whose chest has been left open, back to the ICU. Julie Tome handles anesthesia.

Back in the O.R. lounge, Jonathan is devouring a big, greasy cheeseburger. When Frank and Mike walk in and see him there, Frank, as always, stays neutral. Mike doesn't know exactly what's happening at this point. He's a little out of focus; the day has been so long and strange and bad. Once again he tells Jonathan that no one'll be getting a heart tonight, and Jonathan "wigs out," in Fackelmann's words. George the perfusionist—who, like Fackelmann, has young children—sides with Mike; he doesn't want to go, either. Debbie Seim is in the lounge, too, when the altercation occurs. Jonathan, still chewing, says, If you don't go, *I'll* go get the heart. Mike argues that nobody can fly out in this weather. Jonathan, who knows that a plane has already been arranged, seems to think Mike is defying him, saying that he *personally* will not go; he senses a gathering mutiny. Jonathan has always told Mike that he'd never put a team in jeopardy to get a heart, and Mike can't believe he's going to go back on his word now. But Jonathan is evidently so worked up over this heart that he asks Mike if he'd like to step outside—actually uses those very words, as if they were in a bar somewhere and not in the O.R. lounge of a major metropolitan hospital. ("I guess he wanted to kick my ass," Fackelmann will say later.) Frank just sits and watches the craziness of it all; nothing surprises him anymore. Jonathan leaves, and then Frank finally breaks the stalemate by making Mike, in Debbie's description, "feel like a weenie."

Come on, Mike, Frank says. If we don't go, the kid's going to die.

Mike is silent for a moment, and then he says, If you're going, I'm going.

Mike goes in search of Roger, who's in his office. In Mike's view Roger is "a mess," completely wiped out from this operation. "He gets real emotional and down on himself when things don't go well," Fackelmann observes. Jonathan has already told him about their contretemps, and Roger advises Mike to "get this squared away between you two." Mike does. He steps into Jonathan's small office, which is immediately next to Roger's, and says to him, If planes are going out, I'm gonna go. I never meant to say I was refusing to go. The exchange is so emotional that it ends with Jonathan, perhaps realizing he was out of line, actually giving Mike a hug, and Mike heading for his locker to grab his coat. The harvest team will soon be on its way to Georgia.

I run into Fackelmann outside the O.R. lounge. He's wearing his green winter jacket with the fancy Cleveland Clinic patch on the left breast. He looks neat as always, with his receding sandy-blond hair combed back, his small trim mustache, and a posture to balance books on. But I can tell he's beat, and his eyes are bloodshot behind the small rectangular frames. He collects the Playmate cooler, a bright-red Igloo that he'll put the heart in—assuming that everything goes as planned. "I'm better now," he tells me. "I just needed to get out. I'm OK."

He's been in the O.R. since eight this morning, it's now after ten at night, and when you're in the O.R. all day like that, between thirteen and fourteen hours today, scrubbed in on two difficult cases, two babies, and this last one taking so long and ending so badly, your emotions rise very close to the surface, he says. You lose perspective and focus. He was preparing now to climb into an ambulance and zip through the snowbound streets to an airport where he'd board a private jet with Frank and George, the perfusionist, and fly to Georgia to race in another ambulance to a new O.R. to assist Frank in taking a live heart out of a dead baby. He loves his job, but it's been a long day. Maybe he'll be able to sleep a little on the plane—if they can *get* to the plane.

Frank stops by the ICU before he hustles out to join Fackelmann. Marc Harrison is there; he'll be looking after Drew for the next six hours or so, till they're ready for him again in the O.R. Frank says to Marc, "Just keep him alive."

The team is leaving immediately because the storm has broken for it. Angie and Bart, in the parents' lounge, have been watching the weather outside. When they realized that tonight was the night

Drew was going to get a new heart, after four months of illness, the family headed back by shuttle bus to the Ronald McDonald House, where they have rooms, to gather clothes, sleeping bags, and blankets, intending to camp out in the hospital. When they left, lightning cracked through the sky, an eerie flash in the snowy night.

But now the sky has cleared. The snow will hold off for about an hour before beginning again. This is part of why Angie says Drew's new heart is a gift from God. The break in the weather will last just long enough for the transport team to take off safely, to fly up through the storm that is so big it has closed airports not only in Cleveland but also in two neighboring states.

At six A.M. Roger leaves his office to join Jonathan in the O.R. He was able to get to the airport last night in his little all-wheel-drive Subaru Impreza, retrieve his youngest daughter, and head home again, eastward. Luckily, theirs was the last exit open, Interstate 271 having been closed just beyond it. Roger turned left between the two stone pillars marking his driveway after two A.M.; by three he was in bed, having set the alarm to give him an hour's sleep. At four he bundled up, crawled back into his car, and, weaving in and out among abandoned vehicles, drove to the hospital through snow-covered streets, making it into the ICU by four-thirty to help move baby Drew back to the O.R. and then "get Jono out of bed, give 'im a kick in the crotch to get him going," he says. He chuckles, his spirits revived by the rest.

He scrubs and enters Jonathan's O.R. wearing his glasses but not his headlight, because Jonathan is the lead surgeon on this case. Roger is assisting at Angie's request.

Roger says, "Are you fibrillating?"

"No, I'm giving CPR," Jonathan says.

Drew's heart is about to stop for the last time. Jonathan is manually pushing blood out of it to perfuse his brain and body; they'll try to get him on pump quickly now to avoid too much CPR.

Mark Myer, the nurse who helped take Drew from the ICU to the O.R. yesterday afternoon, and tried to comfort Angie, arrives to begin his day as the circulating nurse. The on-call nurse he's replacing begins to go over the case but then asks, "Do you know this child?"

Mark says, "I believe I do."

"Cross-clamp time was three-forty," she notes, referring to the hour and minute when Frank stopped the donor heart in Georgia.

"Has the plane landed?" Jonathan asks.

Bob Cherpak, the scrub nurse, says, "They said they'd call when they landed. They haven't called." Mac and Charlie are scrubbed in to assist Roger and Jonathan; Kevin Baird is on perfusion. Julie Tome is the anesthesiologist, but she'll soon be relieved by Emad Mossad. Debbie Seim has arrived and begins to organize the other items on today's schedule. She's at the front desk when Carlos, the anesthesiology fellow, gets in and says good morning. Debbie tells him what's happened.

Carlos says, "This baby that I was holding has lovely blue eyes. I hope he makes it."

"He's got interesting karma," Debbie says.

Roger and Jonathan can only wait now. Frank has just called: the plane has had to land at the International Expo Center because of snow and congestion at Cleveland Hopkins Airport, but they're on their way. The ambulance driver is apparently insane, and Frank's worried that the siren-blaring vehicle will go flying off the road—it's the scariest part yet in this harvest, which otherwise has been uneventful.

"Can I advocate something?" Roger, the assistant, asks Jonathan. "While we're sitting here, can we remove the pericardial patch so that we have it for the pulmonary artery if we need it?" It's not really a question.

Debbie approaches the table. "Dr. Mee?" she says. She needs to know if she should keep any surgeries on the schedule today. Jonathan, for his part, won't do any other major cases. ("I will not put kids in jeopardy that way," he says. "I have nothing to prove.")

Roger looks up at her and asks, "Can we do the tet?"

Debbie's eyebrows rise, and she chuckles. "We can *do* anything *you* want."

He nods and says, "Let's do that, then"—a complete repair of tetralogy of Fallot. Debbie schedules the surgery for after the transplant.

At 7:30 A.M. the O.R. door whooshes open. It's Frank Moga, still wearing his skintight navy-blue arctic wear over his scrubs, grinning as ever. He squats and gently pushes the bright-red Igloo over the threshold of the O.R.

"It's a beautiful heart," he says, and the door slides shut again.

Delivery completed, Frank returns to the surgical lounge, where Mike Fackelmann is sitting and talking with Debbie. Mike looks unburdened and happy. "The kid had means of forty," he says. "His liver was out in fifteen minutes." This is by way of describing how well everything went. When they pleged the heart, it stopped on a dime, he says—another excellent sign.

"How did the baby die?" Debbie asks.

Frank says, "Circ arrest. SIDS baby."

"Oh, God," says Debbie.

"But it's a *beautiful* heart," Frank says. "A little thing."

Soon Frank and Mike will each try to lie down for a while—they'll be scrubbing in later, on the tet repair Roger has asked to do. Tomorrow is Thanksgiving, with no cases scheduled.

As usual, someone has brought in the morning *Plain Dealer*, sections of which are now scattered over the two tables in the small, square room. On page 1, James F. Sweeney has written in the off-lede, "Interstates and city streets alike were choked to a standstill while trapped commuters were treated to flashes of lightning and peals of thunder between the snowflakes. The gridlock began as early as 4 P.M. and lasted in some cases six hours, extending beyond the usual downtown streets to east of University Circle.

"Emergency vehicles found that flashing lights and sirens were unable to clear a path through the jams," the reporter noted, then proceeded to cite obliterated bus schedules, abandoned cars, a ten-car pileup, scores of other accidents, and airwaves jammed by cell-phone calls.

. . .

Maryanne Kichuk has come in early this morning to know how everything is going and to speak with Drew's parents and with Tim and Kelly Hohman. Ashley, to everyone's surprise, has gotten better overnight. She is not going to die, as June Graney and others expected. The change in her antibiotics has apparently taken care of whatever infection must have caused the dangerous hypotension. But Tim and Kelly have by now learned that another baby, scarcely bigger than their Ashley—a baby who arrived only a day ago—has been given a heart. To them that doesn't seem fair. One of their regular nurses is upset about it, too, and has openly wondered what's going on here—this should be Ashley's heart. The nurse's remarks have only fed Tim and Kelly's confusion and distress.

The events of the last twelve hours have been unusual, things have not gone the way Maryanne would have wished. She typically sits down with the parents of children who have been listed or are about to be and describes the situation bluntly.

"It's not like another chronic illness where you say, 'Your kid is going to be very, very sick for a long time, and then your child is going to die,' " she explains to me in her office. "That's not the purpose of transplanting, and it's not the message we deliver."

The message she hopes to convey, she says, is this: " 'We're going to replace your child's heart because your child's heart isn't working well, but it's not like replacing a car engine. We become wedded to you. We become part of your family. You are going to be seeing us forever. For-ever. More up front. Less as time goes by.' "

She tells them that doctors can't predict longevity; everyone is different. But she does offer some statistics. The surgery itself has a very low mortality, dependent mainly on the condition of the patient before transplant. However, she notes that, according to UNOS data, the average half-life of a child who's had a heart transplant is twelve years—meaning that twelve years after surgery, half of all transplant patients have died. Some patients need a second new heart within a short time; others do beautifully with their first one. The first pediatric heart transplant was performed where Maryanne trained, at Columbia Presbyterian Medical Center in New York, in 1982. That patient, a five-year-old child born with tetralogy of Fallot, is still alive today.

What finally allowed transplantation to become a more viable option, especially for children, was the introduction in 1981 of cyclosporin, a powerful medicine that reduces the number of cells in the blood that fight foreign matter, but that doesn't have the side effects that heavy doses of steroids and other antirejection meds do. Rejection is the first major danger following a transplant, the second is coronary artery disease, and the third is cancer. From the time of the first adult heart transplant, in 1968, until the advent of cyclosporin, long-term results were abysmal. But even this advance is a double-edged sword: cyclosporin is so effective at immunosuppression, says Roger Mee, "that being treated with it is a little bit like having AIDS."

"The point of transplantation," Maryanne continues, "is not to keep people alive. It's to give them a better quality of life: keep them alive—out of *here*."

Maryanne says that of the more than 4,000 heart transplants performed each year, about 10 percent are pediatric. Organ availability is a critical issue. At any given time more than 75,000 people are awaiting some form of transplantation, double the number of five years ago—even as the number of donors has *decreased,* from 5,000 to 3,500. The Cleveland Clinic has a large adult transplant unit and a medium-size pediatric program. In 1998 the clinic did a total of 113 transplants, the most ever performed at any institution in the United States in one year (the previous record of 108 had been held by Columbia Presbyterian). It will close out this year having transplanted 76 hearts, 12 of them into pediatric patients, the eleventh of whom is currently on the table. By comparison, Maryanne's former employer, Columbia Presbyterian, does about 20 pediatric hearts annually.

"The patients who do the best," Maryanne tells me, "are the ones who believe that everything is going to be OK. I've seen kids die who shouldn't have, because they lost hope.

"You've got to have a lot of stamina, but the single thing that keeps these kids alive is their will to live. Without a doubt. It's really hard to kill somebody who wants to stay alive. It's really hard. And I say 'kill' somebody because I think a lot of what we impose on them is the stereotypical 'Oh my god, a *healthy* person wouldn't survive this.' It's almost counterintuitive. There are patients who get better *in spite* of what we do. And there are patients for whom everything can go perfectly, and they're still debilitated, and they die. I don't know if it's chemical. I think it's more spiritual. But it's definitely there."

Maryanne illustrates her belief in an actual "will to live" by recalling a case from the mid-1980s, an adolescent boy who had had numerous heart operations, as well as cancer, and had survived them all, until his Fontan began to fail. (A Fontan is a repair that allows a heart with only one ventricle to pump to the whole body; when one fails, as sometimes happens, the result is a slow, miserable death.) He was tired. He didn't want a transplant; he'd had enough. But his mom—she was a single parent; the father had died years earlier—talked him into it, and he agreed grudgingly to be transplanted. "When we got the heart for him," Maryanne says, "and the O.R. techs came to take him into the operating room—it was the *Star Wars* days, and he had one of those giant light-stick things—he fought them off with this light stick. He did not want to get on that stretcher.

"And he came through the operation beautifully. The heart was very good—short ischemic time, wasn't out of the body very long—and he required little support. And he came out"—here Maryanne's voice goes high-pitched and quiet, as if she were about to cry—"and he just started to *die*. And there was nothing we could do about it. Nothing. He just died. He had massive rejection, everything got infected, he just died.

"There are kids—look at Jessica." It happened: Jonathan transplanted her two weeks ago (for the second time), after Appachi had resuscitated her with chest compressions, and Jonathan himself had put her on ECMO and told her mother two days. This nine-year-old girl who one doctor thought was a hopeless case got a new heart. And she came through. It turned out she had a growth on her liver—cancer is a huge problem for transplant patients—and had to have surgery for that, too. But as Maryanne says, "That kid wants to live. She just *wants* to *live*. She wants to be with her mom, and her sister and her father and her stepfather. She just wants to be here. And no matter what we throw at her, she's going to come through it. No matter what. This poor kid. She looks like she's been in a war. Think about what she survived—an MI [heart attack], being resuscitated, ECMO, all these things—and she just keeps coming back for more. She doesn't want to come back for that liver surgery, but she'll do it.

"Interestingly, babies—this is my whole 'philosophy' of transplantation—I think babies do OK because they don't know anything else other than living. It's very instinctual. They do OK. Poor little Ashley. She's gonna keep going. She hears her mom's voice and she knows she's alive and this is the only thing she knows. I think babies are born with it."

The heart is in a sterile solution inside a plastic bag. The plastic bag sits in a slush of ice water in a small pail in the cooler.

With the new heart in the room, Jonathan removes the old heart, a swollen, corrupt muscle covered with dark-red gristle, spotted black from countless bovie zaps. Jonathan passes this heart to Kale Buckenmeyer, who's taken Bob's place as scrub nurse. Kale sets it on a sterile towel on the setup table, then squirts it with ice water. It lies there on the green towel, a flaccid, dead thing. Kale draws his finger over the right ventricle, and it begins to beat. It pumps several times, lying there connected to no life whatsoever; it beats, then stops forever. Kale smiles and shakes his head. A student of pathology has been sent to fetch it; Kale puts it in a cup and covers it, then the student drops it in a bag and carries Drew's heart away.

His new one lies on green towels in front of Roger. Jonathan is putting the initial stitches in, connecting the back of the heart first. This heart is pale pink, the color of veal, smooth, glossy, glistening. It's about the size of a Thanksgiving turkey's heart. Jonathan places the first stitch into the patch of original heart containing the four pulmonary veins, then connects that stitch to the left atrium of the new heart. With the lungs collapsed, Drew's chest cavity looks completely empty. When Jonathan has gotten stitches into the distal vessels of the chest, he rolls the heart into the empty cavity as if returning a fish to a pond. After a half hour more of stitching, he tells Kevin, the perfusionist, to start rewarming the blood, which has been cooled to 28 degrees Celsius (about 82°F). As soon as Jonathan finishes the anastomosis of the aorta and the pulmonary veins, he can remove the cross clamp, thus allowing blood to flow into the coronary arteries to feed the heart. If all goes well, the heart will begin to beat.

"What's the cross-clamp time?" Jonathan asks Kevin. Kevin glances at the clock, then at his clipboard, and figures it in his head. "Four hours and twenty-one minutes."

When Jonathan removes the cross clamp, blood flows into the heart muscle, which flushes with color and immediately starts beating. Roger Mee has been transplanting hearts for a dozen years now, but when he sees this happen, he shakes his head and says, "It's amazing, isn't it?" The heart transplant is one of the easiest procedures pediatric heart surgeons perform, but it never loses its powerful effect on the imagination. "It's amazing," Roger repeats. "The beating of a healthy heart." It's something he almost never sees anymore.

The atria fibrillate, and blood begins to froth up around the heart. Because Drew has been on cardiopulmonary bypass for so long, his blood has no clotting factors, but the team has anticipated bleeding. There are no surprises. Soon the heart picks up its ceaseless rhythm, and the procedure continues routinely.

Roger Mee sprawls in a cushioned chair in the O.R. lounge. He's so far down in the seat he's nearly horizontal. He looks tired but in good spirits. He's telling stories about operating in Egypt and Saudi Arabia. Maryanne arrives, the first chance she's had to see him. She hasn't met with the parents yet, preferring to wait till Drew is safe and stable in the ICU (Jonathan's still in there finishing up; Mac and Charlie will close the chest), but she wants to know how they're taking this.

"They seem well motivated," Roger says. "The hubby was involved, asking questions. I told them it wasn't going to be a breeze. When I brought up transplant, the mother started crying."

Maryanne says, "When they cry, they understand."

"And they said, 'Thank you.'"

Maryanne says, "Definitely the hand of God in this one."

When Jonathan arrives, his personality fills the room. He removes his headlight and hood. His hair is matted down from it. He washes his hands, wipes the edges of his mouth with a damp paper towel, then sits and explains how he got the heart, how Smith and another doc gave it up. "So," he says, "we'll have to call them and thank them."

Soon, Roger and Jonathan stand to leave, to speak with the parents. Roger punches the metal plate on the wall that opens the automatic doors and leaves with Jonathan. The woman on the other side of the door is bawling. They pass her and head for the stairs to the third floor. The woman holds a baby boy in her arms. She is inconsolable—weeping loudly. She hands her baby, born ten months ago with tetralogy of Fallot, to Paula Bokesch, the anesthesiologist, who will carry him into the O.R. and put him to sleep, and paralyze him and intubate him, so that Frank and Mike can open his chest and Roger Mee can patch the hole in his heart, resect some of the right ventricle, and enlarge the opening into the main pulmonary artery. The doors close behind Paula, and the mother stands in the corridor, sobbing into her hands.

6. Serious Business

Here was the world of cardiac surgery on babies born with heart defects, and here, I had learned during my first month in this center, was the unusual work of unusual people: Mike Fackelmann and Frank Moga, the ICU intensivists, the cardiac nurses, and Jonathan Drummond-Webb and Roger Mee, surgeons who more than virtually any other kind of doctor determined whether a sick child lived or died. And here, too, was what two parents, at least, could endure on behalf of their baby. Drew H. was a normal child in all respects save for his heart. At some point—about the time Angie learned she was pregnant—the speck of heart cells within the pea-sized embryo had misperformed an important step: the single outflow tract, the developing helix that was the pulmonary artery and aorta, failed to rotate properly as it divided into two arteries, which therefore arose out of the wrong ventricles. Drew's heart at the time was a mere dot, the diameter of a pin, but the shape it took in this second month became an unchangeable error, a defect that would cascade into the horrifying events of his first five months, and his near-death. I had become a spectator of high-stakes games—surgery on infant hearts, the strange ethos of the surgeon and his necessary, double-edged, walk-on-water ego, the PICU that could be quiet as a desert campsite, in one intensivist's analogy, with coyotes forever circling silently just out of view—and I'd gotten, just barely, to know the blunt character of

Roger Mee, who was the focus, the star, of this center, but whose work constituted only a fraction of its whole; and I'd seen that while his talents were excellent on an international and, arguably, a historical, scale, those talents were neither perfect nor unlimited. I would have known all this had Drew not presented at the clinic, but his case added a troubling undercurrent to an already complex and powerful scenario, one I'd had no previous reason to foresee.

The story of this baby circulated quickly through the ICU, from docs to residents to nurses to respiratory therapists and eventually, inevitably, to those parents whose children had had protracted hospital stays and who thus felt a part of the unit. It was one such parent who launched at Angie, Drew's mom, a genuinely cruel comment. I knew this parent to be a kind person, and I honestly don't believe she intended in any premeditated way to cause pain, but she had been brutalized by her own weeks in the PICU—*weeks* during which neither the death nor the life of her only child could be predicted. Perhaps the sharp comment served to ease, if only by a fraction, the searing pain she felt in every moment of her own vigil over her intubated, critically ill infant. It happened on the shuttle from Ronald McDonald House, where both sets of parents had rooms, to the hospital, a casually uttered but blistering remark: *If you had come to Dr. Mee in the first place,* the other mother said to Angie, *maybe you wouldn't have had to go through this.* Angie felt the force of the statement; she felt violated. Who *was* this woman? She and Bart had been in Cleveland for only a few traumatic days and didn't yet know any of the parents here, but this stranger clearly knew her and her son and what his situation was. Angie is a smart, strong soul and knew not to respond, knew also not to dwell on the accusation. What's done is done, she thought; I have my boy and he's alive, and Bart and Drew and I will move through this struggle day by day with our sights set on home, a family of three.

I myself would never have said those words to a parent who'd been through what Angie had. *But I thought it.* And I know I wasn't alone. *Of course* I thought it—what even halfway awake initiate into this world wouldn't have? Doctors here thought about it, too. Here was a not just another "case" but a baby named Drew, born with transposition, who'd needed quick repair after birth—a repair complicated by a funky coronary that would nearly prove to be his demise—and here I was at a pediatric heart center with a man who was, if not the best in the world at doing such repairs, then surely the equal of the best, and the author of papers describing the exact and necessary details for success in fixing numerous unusual coronary formations. Had Mee been Drew's surgeon from the start, extrapolating on his near-perfect switch record, Drew *would* have been at home now, a baby with a prognosis of a more or less close-to-normal life (as far as anyone could predict such things), rather than in the ICU, a new transplant patient with a statistical half-life expectancy of twelve years.

Mee was clearly gifted. Dan Murphy, after moving to a heart center in California, would tell me, "Roger's repairs look like God did them"—in the sense that hearts and vessels looked surprisingly natural when Mee was finished with them, they looked *intended;* there was easy flow, and the angles of the stitched and patched vessels gave the impression of having grown that way on their own. But Mee couldn't do every switch that needed to be done in the United States each year, nor could the small corps of other surgeons who had comparable experience and success with complex switches handle all those hundreds of cases. However, Mee *could* theoretically do, say, the dozen and a half switches required annually by babies born within a 180-mile radius of Cleveland. It's an imperfect world, but in neonatal cardiac surgery, imperfection can have devastating results. Such events as Drew had gone through were all but built into this specialty, perhaps inevitable given its sprawling, self-regulated character. What made the impact even more powerful in this case, more significant, was the fact that if Drew's first repair had gone as it should have, not only would *he* be much better off now, but the other baby who needed that donor heart would have gotten it. The world of congenital heart surgery could seem at times both primitive and unnecessarily dependent on the luck of the draw. My own children weren't born with transposition, but if they had been, I very likely wouldn't have been told about Roger Mee, though he was within walking distance of the hospital where I first held my daughter; and I certainly wouldn't have had the wherewithal to *ask,* nor would I have thought to question our doctor or the hospital. Such hindsight was not beyond Angie and Bart. Shortly after Drew was moved out of the PICU, apparently on the mend, Bart wondered openly if he and Angie had been, to use his word, "buffaloed" at the first hospital. He had just returned to the hospital room with a McDonald's dinner, and said he didn't mean to sound ungrateful—the doctors in Pennsylvania had saved Drew's life, after all. But what he couldn't reconcile in his mind was how they could have sent him home when he wasn't perfectly well, and when "they knew all along" that a surgeon only a few hours' drive away, in Cleveland, *specialized* in fixing heart defects like the one he'd had.

Nothing more was said, or needed to be—as Angie seemed to sense, every ounce of physical and spiritual energy they could marshal must be focused positively on their son now, to ensure his continued recovery. Transplant was forever; as more than one doctor had suggested, transplantation meant exchanging one kind of heart disease for another. Drew, Angie's "miracle boy," was on his way to hospital discharge. And yet for me this case would color everything that would come to pass during my sojourn in the brutal world I was struggling to understand.

Some discrepancy in surgical results was inevitable, given how young this specialty was; it couldn't be understood outside its historical context. Because it was still so new—the first elective closed-heart surgery was performed in 1938, and the first elective open-heart surgery in the early 1950s, though very high mortality persisted until the 1970s and early 1980s—there were bound to be huge variations among surgeons and centers and a lack of protocol, despite the myriad and well-attended conferences and the endless pages of studies and reports published in medical journals. The specialty remained strangely provincial in an era of spectacular worldwide advances in medicine and medical technology. ("My career is built on top of the pioneers, who I think were a different breed of people," Mee told me once, putting himself, a self-described "bean counter" by comparison with those heart-surgery pioneers, into perspective. "They were the cowboys, incredibly bold, to my way of thinking fairly insensitive, but swept along in the excitement, the whole concept of getting a survivor. . . . They had a license to kill because the patients were dying anyway. They weren't adding up their mortalities. As time went by, one center would emerge with a higher percentage of survivors, and they'd all flock over to that center and see what they were doing. That's all stopped. You don't have the prime movers now flocking to each other's places to see what they're doing. Instead you're sitting in your own place, defending what you do and," he adds, chuckling, "being a bit disparaging of what somebody else does.")

The days of the learning curve might be over, but that didn't mean everyone did things the same way and got similar results; procedures and results varied greatly from center to center, with success seeming largely dependent on the surgeon. "In the end, what makes the difference is the surgeons," according to anesthesiologist Paula Bokesch. "You can be the best anesthesiologist in the world, but if the surgeon sucks, you suck." And yet mortality had fallen sharply since the pioneering days—it was astonishingly low, considering what these men and women did: cut open hearts and sew them back together—and for virtually every defect there was now some form of repair. Roger Mee's career bridged those middle years of reduced mortality and the present day of very low mortality but widely divergent success rates among centers. Indeed, in the 1980s he helped to define what could be expected in the way of success at a center performing complex congenital heart surgery on neonates and infants.

Amid the clutter on Roger's office desk on the fourth floor of the M building, the Cleveland Clinic's Children's Hospital, rising above scattered microcassette tapes, sketches on scrap paper of bizarre pulmonary arteries, thoracic journals, a half-full Pepsi bottle, phone messages, stacks of dust-coated financial reports, and wads of chewed Nicorette gum, is a framed portrait of a young woman in her twenties wearing a long dress, simple but stylish, I imagine, for 1930s New Zealand. The woman, seated, hands in her lap, is handsome and solidly built, with strong bare arms and a square jaw.

"When things are bad," Mee says, glancing at this photograph,

"I think of a twenty-eight-year-old woman who went alone to India to start a school for girls. That must have been very difficult in 1936."

Roger Mee speaks of his father, James, a gentle, spiritual man now living in frail health in New Zealand, with limitless love and affection, but he seems to have been made in his mother's image and inherited her frame of mind. Mary Mee was a strong and vigorous woman who, says Helen, "believed in duty with a capital *D*." While Jimmy was busy writing sermons, Mee recalls, Mary ran the house, a vicarage in suburban Auckland. She was an athlete with a degree in mathematics and would be her son's main role model.

She gave birth to Roger, the third of her four children, in Quetta, Pakistan, in 1944, just as the era of heart surgery was beginning (a few months later the third landmark in congenital heart surgery, the Blalock-Taussig shunt procedure, would be performed; open-heart surgery was still nearly a decade away). Mary had gone to Quetta to found an Anglican school and there met her husband, an Irish garrison chaplain. Before the future surgeon was three, a British act of Parliament established the sovereign states of India and Pakistan, ending British rule and sending the region into turmoil. Jimmy and Mary Mee herded their four youngsters into a railway car and fled the Hindu-Muslim strife, returning to Auckland, Mary's home.

Roger was from the beginning a kid who was "always into something," remembers his sister Elizabeth, three years his senior. He was a handful at three, especially difficult on the week's train journey out of Pakistan's chaos, and though basically conscientious and sweet-natured, he remained mischievous throughout his youth. Feeling restless in church, for example, he'd crawl under the pews to tie together the laces of a VIP aide-de-camp's shoes. His father, by then a vicar, discovered that his son's restlessness extended into the evening hours when a policeman knocked on the door to inform him that Roger had been caught shooting out streetlights with a friend. But the incident that was most illuminating of his temperament, according to Elizabeth, happened during one family vacation, when everyone was fishing on a stone wall below a pier.

Roger, then about three and a half, hooked a fish—a big one—which ran, apparently. The little boy either was pulled off the stone wall or fell off it in his excitement at having a fish on the line; in any case, he went under. Knowing that her brother couldn't swim, Elizabeth leaped into the rough current, grabbed him, and dragged him back to the wall, where the others helped them out of the water. Roger had never released his grip on the pole and subsequently succeeded in landing the fish. "He put everything into it," Elizabeth recalls from her home in Kawakawa, an hour south of Auckland.

"He was a bright, brainy chap who was also good with his hands," she continues, citing his scholarship to the university—a necessity, as the family had little money—and his seeming ability to build *anything*. He put an addition on their parents' house and made her a pottery kiln out of an old stove. He was generous, and when he was off at school, he loved to come home to be with his family. "He could do anything he wanted. At school one of his masters wanted him to be an artist. . . . Whatever he tackled, he could do well," declares his proud older sister.

After placing second in the country in national university scholarship exams, Mee breezed through school without doing much studying, preferring sports and fun. Unsure what to do with himself upon graduating, he spoke with his father, who suggested, "Why don't you become a doctor?" Having no alternative ideas, Roger attended Otago Medical School in Dunedin, New Zealand, where he was an unspectacular student who still preferred sports—rugby, cricket, and skiing—as well as making things, goofing off, and drinking. Thinking he might like to be a missionary doctor on the Dark Continent, he also studied, briefly, theology ("Which was *dry—God*," he says now, wincing at the thought of it; the births of his children ultimately canceled any such altruistic notion).

Roger and Helen, who first met at a university orientation dance, were married in 1967, and sensing that as a married man he ought to start making some money, Roger found a night factory job building electric stoves. He slept all morning, copied friends' lecture notes in the afternoon, and returned to the factory each evening. "Our oldest was due on the day of my final examination," Mee says of his last year of med school, 1968. "By then I'd lost interest completely in medicine. . . . Obviously I finished the degree, but I hardly studied."

He would begin his career by moving through traditional medical rotations: cardiology, critical care, the dialysis unit, general medicine. Roger and Helen and their two sons moved twelve times throughout New Zealand during Mee's two-year general residency.

Helen remembers that even then, in his midtwenties, her husband had extraordinary confidence and was always arguing with his superiors, whom he often found to be sloppy, lazy, and slack—though this may simply have been a reflection of his feelings about medicine generally.

"Medicine seemed messy," Mee says. "It was not definitive—a lot of wanking, really. You were treating people who were chronically ill and you got them a bit better and they went out of the hospital, then they'd come back again, a little older, some more systems blown." He chuckles and says, "It was just a feeling that we weren't really *fixing* anything."

He didn't know what to do after this residency, given his distaste for medicine, and so once again he turned to his father, who said, "Why not go into plastic surgery or orthopedic surgery? You could fix things."

Roger followed his dad's advice and gave it a go, doing neuro, orthopedics, plastic surgery, and, he says, "quite a lot of general surgery." He still had no sense of what he wanted to do with his life, but he was happy not to have to worry about it for another four years while he completed his surgical residency.

Then this, too, alas, came to an end. Mee remembers sitting at a table with his fellow surgical registrars, or residents, as Professor Eric Nanson, their adviser, questioned each on his next move. Everyone else knew exactly what he intended to do. When Nanson asked Mee, Mee said, "I have no idea." Nanson insisted that he must have *some* idea. "Not really," Mee recalls saying. "And then I came out with the stupid comment that I was afraid of becoming bored. Nanson's eyes flashed, and he said, 'Why don't you go into cardiac surgery, then?'"

Since he didn't have anything against cardiac surgery, and since the training would conveniently take up another two and a half years in a way that wasn't likely to be boring, he said OK. Nanson helped the thirty-year-old resident secure a place at Green Lane Hospital in Auckland, under chief surgeon Brian Barratt-Boyes. Green Lane happened to be among the most renowned heart centers in the world, and Barratt-Boyes, knighted in 1971, was already a legendary heart surgeon, as famous in his country as Edmund Hillary.

When Helen heard about this, she was upset. As she saw it, Roger always took the most demanding rotations available, the ones that would keep him away from home the most, and now he was choosing what she and the other wives knew was among the most difficult courses of all.

Helen was born on a sheep station beneath the alps on New Zealand's southern island, and she maintains to this day the apparently effortless good looks that seem of a piece with her rustic upbringing. Educated in a one-room schoolhouse, she'd been one of three in her class who went on to university. She'd had it in her mind for these last years that her husband, who'd always liked to have a lot of fun and do things outside, skiing and hunting and motorbiking, would eventually become a general practitioner; they'd buy a lovely farm, she thought, and she'd raise lambs, ride horses, and rear the children while Roger played handyman on the weekends. Instead, the night he returned home to say he might be going into cardiac surgery, she realized that what had seemed merely the inconvenient but temporary necessities of his medical training—his absence, their frequent moves—would be permanent facts of their life. He would *never* be home now, and he'd always be exhausted. When their third child, Michaela, was born, Helen would drive herself to the hospital after the contractions began, stay a week, as was customary ("He used to pop in," she recalls of her husband's attentions, "but he was always tired"), and then drive herself and the baby home.

The Mees bought a house in Auckland in 1974, and the young cardiac resident started at Green Lane. In the mid-1960s, when Barratt-Boyes began making the advances in the field that were to create his legend—most significantly, using a technique called deep hypothermic circulatory arrest, in which young children were iced down to internal body temperatures of 18 degrees Celsius (64.4°F), enabling their delicate organs to tolerate the circulatory arrest then necessary for the repair of many defects—heart surgery was not yet thirty years old, and open-heart surgery had been performed for only a decade.

The era of heart surgery began on August 24, 1938, when a junior surgeon performed an operation that had been expressly forbidden by his boss.

In utero, where the fetal lungs don't have to do the work of taking in oxygen to supply the body, the heart is equipped with a vessel called the ductus arteriosis, which connects the aorta with the pulmonary artery to allow the mixing of red and blue blood. In the great majority of cases, this duct seals itself off naturally after birth, when it's no longer needed. In some infants, however, it remains open, or patent. Patent ductus arteriosis, or PDA, is among the most common of heart defects, occurring in about one out of every two hundred newborns and even more frequently in those born prematurely. When, in an otherwise normal heart, a fat little vessel linking the two main arteries stays open, blood that should be flowing mainly to the body reverses and floods the lungs (where pressure has already dropped, precisely to let blood in). This can result in numerous complications, including pulmonary edema and pulmonary hypertension, which can respectively thicken the pulmonary vessels and choke off the blood supply to the lungs. Smaller ducts, though, are likely to cause fewer problems, or none: someone with a minimal defect may never even know he or she has a PDA.

But more commonly there's damage to the lungs over time, and this was the real horror of the condition before an effective repair was developed: children with a PDA might do well for years, living relatively healthy lives, going to school, growing, and then at some point around age ten or eleven, or later, during high school, just as they were approaching adulthood, they'd gradually get sicker and sicker—with pulmonary vascular obstructive disease, an infection in the duct, an aneurysm—and eventually they'd die.

In the mid-1930s Robert Gross, a thirty-three-year-old surgeon at Children's Hospital in Boston who had been experimenting on dogs' hearts in the pathology department, concluded that the ductus arteriosis could safely be sutured closed, or ligated, in humans. If successful, the procedure would give children who would otherwise die young a normal life. The only obstacle was William Ladd, the hospital's surgeon in chief, who explicitly forbade Gross to attempt it: it was too dangerous to try on a child who wasn't gravely ill, he said, and it wouldn't be ethical to take a child who *wasn't* gravely ill and *operate* on him or her, to attempt an experiment; the child would probably die during surgery that had never before been attempted. This hurdle, however, proved to be easily surmounted: Gross simply waited till his chief had left for a month's summer vacation on Cape Cod. During that month, a seven-and-a-half-year-old girl named Lorraine Sweeney presented with a patent ductus arteriosis, and on August 24, as Ladd was presumably sunning himself on the beach or batting around a tennis ball, Gross took Lorraine into the O.R., opened her chest, and, with her heart beating, sutured the duct shut with number 8 silk. The first elective closed-heart surgery was a success.

When a no doubt tan and rested Ladd returned to work and learned of the unprecedented surgery, he was irate. Gross subsequently left the hospital of his own accord, certain he would be dismissed for his willful insubordination. Several months later, though, the hospital called and asked him to come back: evidently he was too good a surgeon for Children's to lose—talent can be an effective form of job security—though Ladd would never really forgive him.

Because excellent results are impossible to argue with, Gross's legend grew. He'd begun to walk on water and reportedly inspired legions of med students to become heart surgeons. Never mind that his famous operation was an experiment, or that he could have killed a little girl who was in no immediate danger of dying; this was the dawn of heart surgery. *Someone* had to be the first to submit to the knife, and someone else had to be so incredibly sure of himself as to believe that performing the surgery was worth the risk of killing a child and torpedoing his own career. As Gross himself would later put it, "If the surgery had failed, my career would have been over. I'd have been a farmer."*

This was not the first heart operation ever. In 1896 a German surgeon named Ludwig Rehn had opened the chest of a twenty-two-year-old stabbing victim, found a hole in the right ventricle, put three silk stitches through it, and watched as the man's heartbeat returned to normal. Others before him had attempted to save people with serious heart wounds, but all had failed. Rehn's first-ever successful heart surgery would remain the only one of its kind until several years later, when an American named Luther Hill would open the chest of a thirteen-year-old boy who'd been stabbed numerous times and save his life by sewing his lacerated heart back together (on a kitchen table, by candlelight, according to Stephen Klaidman's book *Saving the Heart*).

Five years after Gross's PDA ligation, a Swedish surgeon named Clarence Crafoord successfully repaired a coarctation of the aorta—a defect in which the descending aorta is pinched and narrowed very close to where the ductus arteriosis once entered it. Gross himself had been practicing the same repair on dogs but had yet to attempt it on a human. Crafoord, who had likewise experimented on dogs and would go on to be one of the first surgeons to work with cross circulation, had visited Gross to observe his canine research and watched him repair a coarct on a dog. On his return to Sweden in October 1944, he did the first such repair on a human. It was a second milestone. Gross seethed.

The third milestone came just a month later. It seems less dramatic in retrospect, but it would have enormous ramifications for kids with heart defects and is still in use today in various forms. Helen Taussig, a cardiologist and the head of the cardiac center at the Harriet Lane Home at Johns Hopkins, in Baltimore, had noticed that babies with certain defects, such as tetralogy of Fallot,

---

*The story of Gross's famous surgery, and the risks he took in performing it, have been described in numerous sources. For me they are best evoked in *Dr. Folkman's War*, by Robert Cooke (New York: Random House, 2001), where this quote by Gross appears.

turned blue at a predictable time: when the ductus closed off. Perhaps, she thought, heart doctors could prevent babies from turning blue by fitting them with an artificial ductus. She knew that Gross had worked with shunts—a shunt being a passage through which fluid moves, specifically referring to an abnormal flow through a septal defect or to a tube inserted surgically. The story (disputed in some camps) goes that Taussig, a Boston girl whose father was a well-known professor of economics at Harvard, traveled to Boston, found Gross, and asked him, "Can you make me a shunt?" Gross liked to recall his response: "Madam, I tie them, I don't make them." Discouraged, she returned to Johns Hopkins, only to learn that a surgeon there, Alfred Blalock, had been trying to induce pulmonary hypertension in dogs by flooding their lungs via a shunt from a branch of the aorta to a pulmonary artery. She asked Blalock the same question she'd asked Gross, and he said, "Let me think about it." Taussig had realized that the reason babies with tetralogy of Fallot, a blockage in the pulmonary circuit, turned blue when the duct closed was that the duct was the primary avenue by which the heart pumped blood to their lungs. If a surgeon could somehow create another such avenue—that is, reroute some blood to the lungs—it seemed logical to suppose that the babies might be able to get enough oxygen into their blood to keep them alive. Blalock agreed to try it.

In November of 1944, Blalock operated on a baby girl with tetralogy of Fallot. As always, he had his right-hand man with him, an African American named Vivien Thomas. It was in fact Thomas, Blalock's lab technician, who'd done all the actual work on the dogs, and it was Thomas (who had always wanted to attend med school but had never been able) who'd taught the surgeon how to do what would become a landmark procedure: following his technician's instructions, Blalock used a shunt, or tube, to connect the end of the baby's left subclavian artery to the side of her left pulmonary artery. The technique worked, though there were complications. It would take some trial and error to perfect it, but in the end it became an unqualified success. True, it didn't *fix* the problem, but it was a remarkably effective palliation, and Blalock and Taussig were widely and publicly praised. The aorta-to-pulmonary shunt and its many modifications are still referred to collectively as the B-T shunt, after the cardiologist and surgeon who developed the procedure.

Those years are now considered the dawn not of *congenital* heart surgery but rather of *all* heart surgery, because at the time all heart surgery addressed congenital defects. Although fifteen years had passed since Werner Forsmann threaded that catheter into his heart and snapped those X rays, the notion of repairing *acquired* heart disease was still decades in the future. The mechanics of acquired heart disease—how coronaries became clogged with plaque, even the exact nature of heart attacks—were not well understood then. It was the plight of desperately ill children and their anxious parents that pushed surgeons such as Gross and Crafoord, and soon John Lewis, Richard Varco, and Walt Lillehei, and then John Gibbon and John Kirklin, to search for safe ways to fix children's hearts, often leaving a trail of dead kids in their wake.

Doctors could, then, by the end of World War II, offer some form of treatment to children born with three common heart defects: PDA, coarctation, and tetralogy of Fallot. During the war Dwight Harken, an army surgeon, had laid the foundation of his own legend by successfully removing shrapnel and bullets from the hearts of 134 soldiers without a single surgical fatality—thus demonstrating just how much monkeying the myocardium, or heart muscle, could tolerate without fibrillating and thus killing the patient. At war's end Harken would return to the Peter Bent Brigham Hospital in Boston, where he would make good progress in mitral valve repair, as would another maverick surgeon, Charles Bailey, in Philadelphia. (Bailey was famous for once scheduling two experimental surgeries in different cities on the same day, knowing that if he killed the first patient—he did, as it turned out—and people heard about it, they'd never let him experiment on the next case, unless he could pull a fast one, which he did. The second patient lived.)

All of these operations were closed-heart surgeries, meaning that the patients' chests were open but the surgeons worked on the *outside* of their hearts. Actually opening a heart to sew a hole shut was still out of the question, partly because no one was sure how a heart would behave if it was cut open, but also because the inside of the organ was filled with blood; if it was cut open, all that blood would spill out, the surgeon wouldn't be able to see what he was doing, and the patient would die. What was needed, then, was some means of emptying the heart while still maintaining blood flow to the body.

At the University of Minnesota in September 1952, John Lewis, assisted by Richard Varco and Walt Lillehei, opened the right atrium of a five-year-old girl and, having cooled her body to a temperature of 28 degrees Celsius and occluded the blood entering her heart through the cavae, sewed shut her atrial septal defect, or ASD. The girl, Jacqueline Johnson, was discharged from the hospital eleven days later, the first patient in history to have had successful open-heart surgery.

Half a year later another peak was scaled, this one perhaps the biggest of all. For nearly two decades John Gibbon, a surgeon in Philadelphia, had been struggling to build a machine that would both pump and oxygenate a patient's blood so that defects inside the heart could be repaired. He had long believed that in such a machine lay the future of heart surgery, and he was not alone in this conviction: others in the United States (notably John Kirklin) and elsewhere were likewise racing to develop the first viable heart-lung apparatus. But in May 1953 it was Gibbon who put an eighteen-year-old girl on his machine, designed and built with the help of a young company named IBM, and successfully closed her ASD. It was now possible for a surgeon to operate *inside* a heart that was empty of blood.*

---

*Another often-overlooked milestone in heart surgery, without which *no* heart-lung machine could have worked, was the discovery of "antithrombin," an anticlotting drug. In 1915 a young man named Jay McLean, who was working as a research assistant at Johns Hopkins before entering medical school, was instructed to examine brain extracts to try to determine what promoted clotting. McLean decided to test other organ extracts as well. He found that one from the liver, heparphosphatide, actually did the opposite: it *prevented* clotting. His boss, a physiologist named William H. Howell, was the one who named the extract heparin and worked on its purification. By 1937 production had begun on clinical-grade heparin, a powerful, widely used anticoagulant essential in the heart-lung bypass process.

As great an advance as it was, however, Gibbon's machine was not immediately embraced. All of his subsequent patients died, and deeply discouraged, he gave up operating on the heart altogether. One problem was that the technology was still not clearly understood, even by Gibbon himself. It was for this reason that Walt Lillehei focused his efforts at the University of Minnesota on perfecting cross circulation, a technique in which another person's heart and lungs were used to oxygenate and pump the patient's blood. In order to repair a VSD, or ventricular septal defect, the most common heart defect, a surgeon needed a bloodless field so he could locate the hole, inevitably obscured behind a mesh of trabeculae. But the repair itself took longer than that for an ASD—in fact, longer than blood could safely be occluded in a cooled body, as Lewis had established conclusively over several attempts (the mortality was 100 percent). Lillehei's first try was only a limited success: he managed to sew shut the VSD, in the heart of a thirteen-month-old boy whose father lay in the same room, circulating and oxygenating his son's blood, and both child and parent survived the operation—but the boy died several days later of pneumonia. Later that same month, March 1954, Lillehei tried the procedure again, this time on a four-year-old girl named Pamela Schmidt, whose blood was likewise circulated by *her* father. This time the operation was a success: the patient was discharged, healed.

Although he regularly suffered setbacks in the form of dead children, Lillehei persevered. He repaired an atrioventricular canal defect later in 1954, a first, and was similarly the first actually to repair (not merely palliate) tetralogy of Fallot, in a ten-year-old boy named Mike Shaw, whose circulation donor—an essential role that nonetheless carried the risk of death or, as one of Lillehei's donors would discover, brain damage—was a twenty-nine-year-old highway worker, a total stranger to the Shaw family.

Lillehei and other early open-heart surgeons soon learned that poking around in the right ventricle with needles and silk had its hazards: holes in heart walls often displaced the Bundle of His and other critical electrical pathways, which, if accidentally cut or squeezed during an operation, could cause heart block, a potentially fatal complication. To counter this risk, Lillehei in 1957 tried sewing electrodes into the heart muscle of a three-year-old girl—a tet repair who'd gotten heart block—to pace, or regulate, her heart. The technique, yet another first, worked. Less than a year after that Lillehei asked his repairman to make him a small, portable pacemaker. The repairman, working in a garage heated by a potbellied stove, carried out the commission, and in 1958 Lillehei implanted this device in another patient with heart block—one more first.

Congenital heart surgery was now starting to roll, with news and medical details of successes spreading, with the technology rapidly advancing, with Lillehei presenting his enviable results in dramatic fashion. In 1955 a British doctor became the first to stop and then restart a heart using a potassium solution (cardioplegia would remain for many years a controversial subject, given the danger of a heart's not restarting). By 1958 John Kirklin, working two hours south of Lillehei, at the Mayo Clinic, would report 245 cases in which he'd employed a modified version of Gibbon's heart-lung machine. That same year Ake Senning, a Swiss surgeon, devised an ingenious repair for transposition of the great arteries, which involved rerouting blood through the atria to opposite ventricles by creating a series of complex muscle flaps, or baffles. This was called an *atrial*—as opposed to arterial—switch. Senning's operation—called a Senning—had the disadvantage of being *so* ingenious that only Senning himself could perform it.

Six years later, in 1964, "Wild Bill" Mustard, in Toronto, achieved the same end result as Senning by using pericardial tissue, a technique that was much easier to teach to other surgeons. Unfortunately, it was common for the atrial-switch patients to die as teenagers or young adults, or sometimes even earlier, as their right ventricles proved incapable of pumping blood to their bodies once those bodies reached their adult size, or as deadly arrhythmias set in.

In 1970 François Fontan of Bordeaux, France, and Guillermo Kreutzer of Buenos Aires independently performed surgeries to bypass the right ventricle altogether, inventing a procedure, to be called a Fontan, that would be used to palliate many different types of single-ventricle defects. The idea of rerouting blue blood returning from the body, and channeling it directly into the pulmonary arteries so it will be passively pushed through the low-pressure circuit of the lungs, remains among the most imaginative concepts in the annals of congenital heart surgery; no one had ever before suggested that one ventricle might be able to pump to both the pulmonary and the systemic circuits.

By the early 1970s, therefore, a little less than two decades after the first open-heart surgery, there was some form of repair or palliative available for most congenital defects. Surgical innovations would continue, though not, now, in the form of actual mechanics but rather in the management of patients, delicate organisms, as technological advances in cardiopulmonary bypass and intensive care marked the next wave of change in congenital heart surgery. At Green Lane Hospital in Auckland, New Zealand, in the late 1960s, Barratt-Boyes had become among the first to operate on infants using the technique known as deep hypothermic circulatory arrest (his modification of the cooling technique combined cardiopulmonary bypass with surface cooling), in which the patient, in preparation for surgery, was chilled in a slush of ice for hours. At 18 degrees Celsius (64.4°F) the body's organs—most important the brain and heart—could survive for many minutes with no circulation at all, meaning that the patient could be taken off the heart-lung machine while his or her heart was emptied, and the surgeon, if he was fast enough and sufficiently skilled, could refashion that heart in a clear field, with no blood and no tubes in the way.

Still the very idea of putting small infants or neonates on any kind of extended bypass remained all but taboo. Common sense, if nothing else, suggested how dangerous that would be: babies' blood cells were delicate, and their organs and tissues were especially vulnerable to perfusion injuries; their blood couldn't be sent through a pump without those delicate cells' being smashed to bits, thus destroying the blood's ability to carry oxygen, fight infection, and clot. Even sending blood through synthetic tubing could make it go haywire. High mortality was inevitable. Since the first bypass operation, this had been accepted wisdom. Babies born with defects that would prove fatal within a couple of years were therefore palliated with closed-heart procedures—B-T shunts if they needed increased flow to the lungs, or pulmonary artery bands if they needed decreased flow (the latter corrective, in which a band was fitted around the main pulmonary artery to restrict blood flow in the case of, say, a large VSD, having been introduced in 1952)—and sent off to an uncertain future.

Barratt-Boyes began to make his mark at a time when such palliative surgeries were accepted practice, simply because babies were thought to be too vulnerable to withstand the insult of cardiopulmonary bypass. Noting that palliations were accompanied by their own set of problems, Barratt-Boyes sensed that more could be done for these patients. So he built on the work of others (as all surgical innovators must do) and was able to achieve unprecedented success in *repairing*—not just palliating—heart defects in small infants, by lowering the babies' core body temperatures to levels that reduced their brains' oxygen requirements. This allowed him to perform open-heart surgery with no bypass during the actual repair, and only limited bypass throughout the entire procedure.

When Mee started working under Barratt-Boyes in 1974, this method was still in use at Green Lane. Roger and his fellow residents would pack a young patient in a slush of ice, then go off to have a cup of tea and wait for the body's temperature to drop from 36 degrees Celsius to 25. This generally took a good hour and a half. Fat necrosis could be a problem, particularly for babies who had a lot of fat. (A scarier possibility was fibrillation—when that happened, Mee recalls, they'd have to rush in and open up the kid quick.) As soon as the patient had reached 25 degrees Celsius, the young surgeons would stub out their cigarettes, gulp down their tea, scrub in, and put the kid on bypass for the final 7 degrees of cooling, to 18 degrees. One of the advantages of surface cooling was that the heart, being at the center of the body, cooled last; the cardiac output thus considerably exceeded the body's declining demands during the procedure. At 18 degrees they took the patient off bypass and removed all the tubes; after Barratt-Boyes repaired the heart lickety-split, the baby was returned to bypass to be rewarmed. Speed was never more critical. No one else—not the team at Boston, not Kirklin at Mayo—was doing such work on such small children, three months and younger.

But by the mid-1970s the nature and role of cardiac surgery were swiftly changing, largely because of two significant developments dating back to 1967.

The first of these, in May of that year, was an important surgery that attempted to repair not a congenital defect but rather an adult case of acquired heart disease. At a modest but respected group practice in Cleveland, an Argentinean surgeon named Rene Favaloro sewed a vein onto a coronary artery to bypass a blockage and deliver oxygenated blood to his patient's heart muscle. This was not the first coronary artery bypass graft (CABG); one had been done in 1962 and another in 1964, but both were isolated cases and not repeated. It took Favaloro and his colleague Mason Sones, a pit bull of a scientist, to recognize the technique's potential and power.

The single critical event that had led to the CABG revolution was an accident. Sones, the head pediatric cardiologist at the Cleveland Clinic, and the man who had built its cath lab in the 1950s, unintentionally shot some radiopaque dye directly into a right coronary of a patient whose aorta was being catheterized. At the time it was common knowledge that replacing blood with dye in a coronary, thereby depriving the heart of oxygen, would send the heart into ventricular fibrillation and kill the patient. And indeed, Sones's patient *did* go into V-fib, but then, after a few frantic moments, the dye cleared and the heart resumed its normal rhythm. Sones would soon prove, through tests on hundreds and hundreds of cases, that enough dye to furnish a clear angiogram could be injected into the coronaries without endangering the patient. Many of these test cases, however, were unaware that they were being experimented on. Although this was not an uncommon practice in those days, Sones was accused at a conference of evincing dastardly ethics—it's easy to make guinea pigs of others, charged his rivals. Incensed, he demanded, on his return to Cleveland, that a colleague put *him* through the experiment, much as the forefather of interventional cardiology, Werner Forsmann, had done before him.

While blocked coronaries could now be definitively and precisely *diagnosed*, a *solution* to the problem would not be found until Favaloro, a cardiac surgeon, began to work with Sones toward a better means of delivering oxygen to the myocardium. Then came the breakthrough: in May 1967 Favaloro grafted a vein onto the blocked right coronary of a fifty-one-year-old woman; eight days later the vein remained wide open and was delivering oxygenated blood to the previously starved muscle. Sones and Favaloro needed considerable tenacity to pursue the coronary bypass graft and perfect it. Formidable cardiologists would denigrate the procedure for years, demanding proof, through further testing, that it was effective. But enthusiastic clinic surgeons were soon doing 3,000 CABGs a year, and their peers at other centers throughout the country were adding thousands more to the total. Now more than 900,000 bypass grafts are performed each year worldwide, 350,000 of them in the United States alone.

This one procedure, more than anything else, made the Cleveland Clinic what it is today, transforming it, as Stephen Klaidman documented in *Saving the Heart*, from a "private [institution] . . . in a run-down neighborhood of a midsize city often parodied for its provincialism" into a "glittering, world-class medical megacenter."

The second milestone of 1967, which occurred in the final month of the year, was so spectacular, so mind-bending, that it provoked nearly as much excitement as the prospect of men walking on the Moon. The procedure made headlines the world over and turned a surgeon into a celebrity of rock-star proportions—despite the fact that the surgery itself, at least for the time being, had no real practical application (not unlike walking on the Moon)—no long-term survival. Nevertheless, it was the ultimate repair, and it was no longer science fiction but instead, now, fact: successful heart transplantation in a human. On December 3 Lillehei protégé Christiaan Barnard removed the failing heart of a patient named Louis Washkansky and plugged a new one into his empty chest.

Everyone within the medical profession knew that the *real* pioneer here was an American surgeon named Norm Shumway, but the press was more interested in the South African surgeon with the Pepsodent smile. "Shumway conceived the idea [of heart transplantation]; this was his life's work," says Boston pathologist Richard Van Praagh, who was thirty-seven at the time. "Our friend Chris Barnard was a cowboy, which is very important in surgery. He told people after visiting Richard Lower, one of Norm Shumway's former colleagues, in Virginia, that he was going to go home and do it [attempt a transplant]. And by God, he did. But he didn't have the experience or the background. The reason Norm Shumway didn't do it was that he knew it was too early yet. He knew they hadn't solved the rejection problem."

Sure enough, Washkansky died seventeen days after the surgery, but by then the headlines were old news, and Barnard walked on water. After a brief rage of popularity, transplantation went dormant for more than a decade, deemed impractical until the advent of the antirejection medication cyclosporin in the 1980s. Most docs today know which man was "the real hero of cardiac transplantation," as Van Praagh puts it: Shumway would ultimately become one of the most prolific and successful heart transplantation surgeons in the country. But Chris Barnard, with his movie-star looks, was launched into international fame by the surgery—a fame that would last until his death, in 2001, and beyond—and he would keep the glory. The world likes "firsts," and the dashing "cowboy" fit the mold.

Not many years later, on the other side of the world, Mee witnessed Barratt-Boyes's reaction to the proliferation of these adult-centered procedures, particularly CABGs. Barratt-Boyes hadn't been quick to embrace the technique, but its undeniable benefits eventually induced him to do some forms of coronary surgery—but "very reluctantly," says Mee. "I think he foresaw that coronary would swamp congenital. Which was exactly what happened, exactly what happened."

Mee's time at Green Lane was a fruitful one, and he met many topflight surgeons who came to visit Barratt-Boyes. "I think I quite enjoyed it," Mee says now of his two and a half years there, from 1974 to 1976. "It was exciting, quite challenging, but it was still just a job. I was much more excited by building a bach down at the beach." When he wasn't working, he joined Helen in this latter project, an A-frame house; Helen was pregnant at the time, he recalls, and mixing the cement in a wheelbarrow. As hard as his cardiac residency was, the work itself still didn't mean all that much to him, not really, though it was thrilling in its way because of all the attention Green Lane received, the famous surgeons who came to observe. And he liked being busy. (When he picked up hepatitis in the hospital, it laid him up for two weeks, during which convalescence, exhausted but bored with lying in bed, he built a bedroom, a carport, and a concrete veranda at his parents' house. The illness put him off beer for two years, he recalls, but not cigarettes—he vowed to keep that hobby strong.) But for all that he considered surgery "just a job," he seems to have begun during this period, for the first time ever, to doubt his ability. He continually asked Barratt-Boyes, "Am I good enough to do heart surgery? Am I good enough?" At the time Barratt-Boyes was noncommittal, saying vaguely that the decision was up to him. Decades later, in 1997, Barratt-Boyes, asked to comment on his former resident, said that Mee's results were now significantly better than his own and that "he is setting an example we are anxious to follow throughout the world." He added that Mee "was not an outstanding trainee."

"I came out of that place," Mee says, "thinking only God could do congenital heart disease. It never crossed my mind to do this."

The mid-1970s were still a time of very high mortality. Infants born with the more severe defects, such as hypoplastic left heart syndrome, were simply left to die, without so much as an attempt being made to repair them. Very small babies couldn't be operated on at all, and parents knew to expect the worst. In those days the quality of a repair and the quality of life following surgery were not even discussed; the goal was merely for the patient to survive the operation, make it out of the O.R. This situation began to change in 1972, after Robert Gross, the founding father of pediatric heart surgery, retired from Children's Hospital in Boston. His replacement was a surgeon named Aldo Castañeda.

"My belief in myself, my fire, got lit when I got to Boston," Mee explains.

Before the end of his residency at Green Lane, Mee won a fellowship that would pay travel expenses to Boston so he could work at the esteemed Peter Bent Brigham Hospital. Helen was game to make the move to the United States; they packed up their belongings and their three children, aged one through seven, and boarded a plane in Auckland. After the family made its way to temporary lodgings and deposited the bags—it'd been a murderously long trip with the kids and the piles of luggage—Mee decided to pop into his new hospital to say hello and introduce himself. The moment he did he was told, "You're wanted in the operating room."

"I didn't get home for three weeks," Mee says.

As it turned out, the Brigham had initiated a program that would award a chief residency to one of its fellows. Because of a glitch in the accreditation for training in this position, it was left vacant, and Mee got it by default. At thirty-two, in what was more or less a fluke, he found himself suddenly the chief resident in the division of thoracic and cardiac surgery at a prestigious American hospital, as well as a clinical fellow at a well-regarded American medical school, Harvard.

If Helen had thought, back in New Zealand, that her husband was neglecting his home life for work, she would soon be exposed to a more American-style work ethic. "She was going to leave me, basically," Mee says now. "And I think that made her, because she developed some self-determination. But God, that first year was really tough. I was on duty every night, seven days a week. I always tried to get home, even if it was just to have a shower and have breakfast and go in again." Helen bought them an old house in Wellesley, a fixer-upper. Occasionally he'd get a day and a half off, which he and Helen would spend tearing out a wall in the house, laying a patio, or rebuilding a fireplace.

Boston proved to be a place of book-smart doctors. "They had me snowed," Mee says. "But what I found was that I had better common sense than they did, and better judgment, could put things together better. So my confidence started to grow."

Helen remembers it a little differently. He stopped having fun, she says; he changed. He realized that this was serious business, that he had to work all the time, and, moreover, that he liked the work. "He went up like a star in Boston," she says. "He disappeared. He lost all his hair."

She recalls that during that first year he had just two long weekends off. On the first of those they traveled to Toronto to visit some old friends; on the second, those same friends came to Boston. Other than that, the only time she saw her husband was when he came home for breakfast (to this day they have breakfast together every morning, even if he has to leave before dawn to do a transplant). But those years were pivotal for Helen as well: for the first time in her life, she explains, she was around wives who didn't just talk about their husbands all day, who actually had lives of their own.

Because he was chief resident, Mee was obligated to rotate through the cardiovascular surgery unit of Children's Hospital, which was situated directly behind the Brigham. Had he been only a fellow, as he'd expected, he'd likely have stayed in adult cardiac surgery; instead, by chance, he spent six months as chief resident under Aldo Castañeda.

Castañeda had arrived in Boston five years earlier from Minnesota, where he'd done his residency under the Lillihei regime and then become a staffer. Even as the University of Minnesota sank from view in the world of cardiac advances, Castañeda would carry on its tradition of daring, even harrowing innovations in cardiac care at Children's, pushing congenital heart surgery into a new era by operating on smaller and smaller infants and ultimately performing the first-ever elective neonatal open-heart surgery.

Born in Genoa, Italy, in 1930 to a Guatemalan physician father and a mother of South American–Austrian descent, Castañeda grew up in wartime Germany, studied in Switzerland, moved to Guatemala, and by 1958 found his way to the University of Minnesota, despite the fact that English was not yet one of the many languages he was fluent in. A tall, elegant man with white hair and a high, nasal accent of indeterminate European origin, Castañeda today has forgotten little about his early training or what prompted him to attempt open-heart surgery on neonates. "Even when I was a resident in Minnesota," he says, "I saw those kids come in, and two things I figured out: first, some came too late, and their hearts were damaged or, more important, the lungs were irreversibly damaged; and second, I saw some of the consequences of these palliative operations in children who came later—either they had a very distorted pulmonary artery at the site of the shunt, or the shunt had been too big and they had pulmonary vascular obstructive disease."

Dissatisfied with the long-term effects of palliative procedures and believing there was no reason infants and very young children could not be operated on successfully, he and his colleagues at Boston began repairing tets and VSDs in neonates and infants in 1973, advocating strongly for early and complete repair. For the remainder of that decade Children's in Boston, Green Lane in New Zealand, and the University of California in San Francisco, under chief surgeon Paul Ebert, were the only centers in the world endorsing and offering early repair utilizing Barratt-Boyes's advances in deep cooling and circulatory arrest. These institutions' three senior surgeons, according to some sources, worked so quickly and with such masterly skill that they were capable of carrying out complex repairs within the time constraints imposed by circulatory arrest; halting a baby's circulation is never entirely without risk, but the chances of irreparable brain damage increase drastically, it was then believed, after sixty minutes of circ arrest (thirty minutes is now considered to be the comfortable limit, according to Mee).

The 1980s brought advances at all levels, from improvements in the size and material of surgical instruments to refinements in cardiopulmonary bypass to better postoperative care, but most important, according to Castañeda, was that accumulated experience with deep hypothermia made early repair safer and safer for smaller patients. There remained, however, one group of patients for whom there was no good palliation: babies with transposition of the great arteries.

In 1975 a Brazilian surgeon named Adib Jatene had performed the first successful arterial switch on a young patient born with transposition of the great arteries. Until then every surgeon who'd tried the procedure had been unable to get the coronaries to work after switching the arteries, and so had been forced to compromise by redirecting the blood from the atria. Although Jatene's arterial switch operation was the first, it would be an Egyptian surgeon named Magdi Yacoub (today practicing in London) who would popularize the procedure. There was just one problem: Jatene's switch could be done only on older children.

Castañeda was interested in these patients, children with transposition who would die without early intervention. Why *couldn't* they be repaired in the neonatal stage? He knew that at birth the ventricles were virtually identical in mass, then rapidly changed size according to their intended jobs. The current procedures, the Senning and the Mustard, which had barely any mortality at Castañeda's center, were good palliatives in that they allowed the heart to pump blue blood to the lungs and red blood to the body, but they weren't *repairs:* the arteries remained transposed, and the kids who underwent these surgeries would often start to fail from arrhythmias and right ventricular failure as teenagers and young adults. If they didn't die on the operating table, neither were they expected to live long lives.

Castañeda thought modern medicine could do better for such kids. The arterial switch had been around for eight years now, and Castañeda reasoned that if he could perform such a switch while the ventricles were still of similar size, the heart of a baby with TGA might develop normally—without arrhythmias by age ten, without abnormal ventricular development, without the likelihood of heart failure in early adulthood. But it had never been done successfully this way before, and his hospital already offered procedures that palliated the defect with nearly zero percent mortality. What parent would choose for his or her child an operation that carried a high risk of death, over a different procedure with almost none at all?

"The father of the first patient was an engineer," Castañeda remembers. "I explained [the theory] to him, and he thought it was a good idea. A higher risk, but for the future, he thought, if you could have the left ventricle as the systemic ventricle, that was obviously an advantage." Castañeda performed the first neonatal switch on January 2, 1983. "We did it," he says, "and fortunately it went very well." He completed his first fourteen neonatal switches with just one death. "The mortality very soon was very low," he notes— "surprisingly good results. . . . That of course encouraged us to tackle other lesions—truncus arteriosus, interrupted aortic arch— and we did them all as neonates, as soon as possible."

During these years Castañeda was well known and drew legions of visitors to Children's. He was transforming the practice of congenital heart surgery in two critical ways, by encouraging early and complete repair and, no less significantly, by arguing for separate services for surgery on congenital and acquired heart diseases. Cardiac surgery on adults, surgery to fix the number one cause of death in the United States, was snowballing. In most centers cardiac surgeons now worked mainly on adults and did kids almost on the side; it was a simple matter of numbers. Castañeda was instrumental in creating a congenital-heart-center mentality, getting hospital boards and other doctors to acknowledge not only that congenital defects were so complex and varied as to require their own surgical subspecialty, but also that neonates and babies and children needed different kinds of management, relative to their size and weight, in preoperative care, anesthesiology, cardiopulmonary bypass, and postoperative care.

In January 1977 Mee arrived for his first day at Castañeda's congenital heart center. Those were the days, Mee says wistfully, when a pack of smokes and a Coke would do him quite nicely till 10:30 A.M. Operating under Dr. Castañeda was different from anything he'd previously experienced. Mee was familiar with congenital from his residency under Barratt-Boyes, but he hadn't been the one doing the surgery at Green Lane. Here he would be. Castañeda had so many visitors and was so interested in teaching that he had what amounted to bleachers in the O.R. Because he did so much talking, and because Children's was a hospital devoted to teaching, he didn't perform the less complex operations; in these he would assist Mee, his thirty-two-year-old resident. As Mee remembers it, this made for some difficult cases. Castañeda would get so carried away telling the audience about the particulars of his method for repairing, say, a ventricular septal defect that he would forget to assist. Finally Mee had to have a talk with him in his office. "I can't operate this way," he told his boss. The next day when Castañeda carried on exactly as before, Mee got so fed up, he says, that he'd have set down his instruments and left the table in the middle of the case had he not been roped in on either side by the bypass tubing.

Mee nominally performed about a third of the heart operations Children's did during those six months, but he remained unsure of himself, unconvinced of his abilities. Castañeda, who would open a congenital heart center in Guatemala in 1997 and today, at the age of seventy-two, continues to practice and teach, occasionally passes through Cleveland. At dinner one evening not long after Drew H.'s transplant, he was amazed to hear Mee admit to being uncertain of his skills as a heart surgeon. "You really thought that?" Castañeda asked, then paused and added, "Actually, that's how I thought *I* was, too." Castañeda says his young protégé was an excellent surgeon even then, but Mee shakes his head. Van Praagh confirms it, however, saying of the then thirty-two-year-old Mee, "He was a superstar when he was here."

But that six-month stint in congenital was soon over, and then it was back to the Brigham as a junior staff member, where Mee started doing the acute add-on cases not performed by his superiors, Jack Collins and Larry Cohn—including the unstable left main coronaries, where just getting the patient *into* the O.R. before he or she had a heart attack and died was a nail-biter. And even then he couldn't have the O.R. until the big guys were finished with their scheduled cases, so he never started his till six at night. He loved the challenge, though, of getting those unstable anginas and left mains on bypass, and began to learn the delicate manipulations of anesthesiology there. He would do two full years in adult cardio-thoracic surgery at the Brigham, following his half year with Castañeda. Toward the end of his three-year sojourn in Boston, the call came: it was D'Arcy Sutherland in Australia, asking if he'd consider taking over the struggling congenital heart center at the Royal Children's Hospital in Melbourne. Mee immediately thought, No, I do adult, not congenital. And yet in the back of his mind he knew that 80 percent of his work, if he stayed on his current course, was going to be coronaries. And that was going to get boring fast.

"I wasn't consciously thinking about this," Mee says now, "but there was no question that congenital was so varied, required so much more thinking individually with each patient, that when D'Arcy Sutherland asked if we'd come, I said, 'I don't do peds,' but that's when I started thinking."

After three solid years of surgery Mee had realized, he says, that "I might be able to be a heart surgeon." In Boston he'd begun to change from, in his word, a "slapdash" kind of guy to an obsessive *i*-dotter. For a New Zealander who'd previously considered obsessiveness a character flaw, this was no small transformation, but it was one that's probably not uncommon in the high-stakes world of cardiac surgery. It's one thing to be a less-than-obsessive carpenter, say, or cook, or accountant—you're not going to kill anyone if you screw up. But if you're a cardiac surgeon, people live or die by your hands, so if you can decrease the number of patients who die by becoming obsessive about the details, you're going to— if you're a *good* surgeon—become obsessive about the details.

Such obsession can't translate to all areas of the surgeon's life, though. Mee has had to bite his tongue at home for the sake of domestic harmony, knowing that to press his O.R. expectations onto the family routine could be harmful.

Although Mee has spent the majority of his waking adult life away from her and their kids, Helen says she has never doubted his need for his family, or his loyalty to it. Nor, even in the most stressful times, did Roger fail to appreciate the pressures on Helen or her huge workload—no small matter in a profession that leads many couples toward divorce. Before he completed his three years at the Brigham, he had nine thousand dollars in grant money left over and asked Collins, his boss, for a few months' leave for travel. Instead of leaving his family in Boston while he pursued his plan of visiting as many American hospitals as he could before heading off to start his next job, he bought an old motor home so they could all spend three months together crisscrossing the continent, stopping wherever there was a center Roger wanted to see—Quebec, Montreal, Buffalo, Cleveland, Milwaukee, Chicago, Minnesota, Vancouver, Seattle, Portland, Sacramento, UCSF, Stanford (where he photographed a transplant with a movie camera that took single images), UCLA, San Diego, Houston (where Denton Cooley did thirty-one cases on the day of his visit), New Orleans, and Birmingham—and then driving up the coast to Baltimore and New York and finally back to Boston, where he completed his work and the family prepared to leave. Mee told me that he was well known for not going to bed with the nurses—his is a profession in which such dallying is not uncommon—and so, as going-away gifts, the nurses at Children's gave him stuffed sheep, in honor of his native land and its inhabitants' reported proclivities.

The Mees landed in Australia in the summer of 1979, Roger having in the end agreed to Sutherland's request that he take on the directorship of a congenital unit that had already seen five surgeons sacked for bad results. Mee had told Sutherland, "I only did six months [in congenital], two years ago." Sutherland said it didn't matter; he'd talked to Castañeda and Collins, and that was all he needed—he wanted him for the job.

Helen still recalls Roger's six months at Children's this way: "He loved it—probably because he got to operate so much." Mee was named director-elect of the department of cardiac surgery at the Royal Children's Hospital in Melbourne, beginning in July, shortly before his thirty-fifth birthday.

The Royal Children's Hospital, in the heart of one of Australia's capital cities, was the biggest children's hospital in the Southern Hemisphere, with 460 beds. While it was a prestigious institution generally, its cardiac unit "had fallen on hard times," according to Sutherland, who had been recruited to turn it around, though he was sixty-six years old then and on the verge of retiring. When Mee arrived to replace a surgeon who had been, in Sutherland's description, "totally inadequate," he found its equipment antiquated, its care lax, and its cath lab a disgrace, with machines held together by duct tape. The O.R. and ICU staffs were cynical and dispirited, having watched seven of every ten neonates who were operated on at the center die as five surgeons came and went. To make matters worse, the previous surgeon had also come there from Boston, so the staff wasn't inclined to think much of Mee when he turned up. Why should he be any different, this chain-smoking kid who stood just over five feet five?

But Mee put his head down and went to work, unconcerned about salary, about what people said about him, about contracts and provisions, about the hospital administration's reluctance to make funds available for new equipment. The place was a mess, and no one, save D'Arcy Sutherland, was going to be of any help. "It was very hard coming into a system that was so passive," Mee explains.

"I was going to bring them kicking and screaming into the twentieth century whether they liked it or not," he goes on. "It was fight fight fight the whole damn time. I was in a state of anger for most of the first ten years." Although the hospital worked within the country's system of socialized medicine, it still took private patients who could pay. Mee chose to be on a part-time salary, but his private cases were relatively lucrative. Because the administration could not allow him to purchase new equipment, provide travel money so he could attend seminars, or hire surgical fellows, he found ways around these obstacles; out of his private-practice money he paid himself what would have been his full-time salary at the hospital (about $180,000 Australian, he says), then used the rest to help set up a fund to buy not just state-of-the-art equipment but basic stuff such as a flash sterilizer for the O.R. instruments that the hospital refused to purchase, as well as to travel to congenital conferences and to hire fellows.

"Bets were on I'd only last a year," Mee recalls, "so I said, Fuck it." He'd *make* things work by any means possible. Happily, the cath lab, which Mee says was inferior even to the animal cath lab he'd worked in at Green Lane, blew up, forcing the hospital to update the equipment.

As head of the husbandless household, Helen was aware of such matters as income and expenses, and she remembers that her husband gave up (to his slush fund) as much as 50 percent of what would have been his annual pay. She remembers being angry, too, during these years, not at the hospital administration, and not because of Roger's income (which was more than comfortable, reduced though it was), but rather because she almost never saw him, and they rarely talked. And when they did, it was often charged with his anger at the hospital. So not only was he working harder than he ever had, the work felt not satisfying but frustrating. Whenever he tried to spend time with the kids, he would fall asleep out of exhaustion. She remembers saying to him in anger, "We're not happy." And he responded, she says, " 'We're not here to be happy.' "

He was soon immersed in a unit where mediocrity was the standard, the administration was unhelpful, and infant mortality was an appalling 63 percent—though that hardly meant anything, because the results were so dismal that everyone who could go elsewhere did: the center did about 250 cases a year, only a handful of those infants. Mee had worked, by little more than chance, under two of the best congenital surgeons in the world, at two of the best centers, where infant mortalities were well under 20 percent. He knew how to do the surgeries, and he knew what level of quality a system and a team must maintain—from equipment to staffing to protocol—in order to achieve those results. (Repeatedly, the surgeons with the best results have stressed to me that what they themselves do is a small part of the work of a big team.)

"I was learning how to turn grays into black and white," Mee says of his first years at the Royal Children's. "Which you have to do to make a decision. To me everything's gray. You have to do what computer enhancement does, you have to say this is a dark gray, so it's black. Then of course you get accused of only seeing in black and white because you're so decisive. Physicians love grays. Let's face it"—he grins—"they're medical people, they love grays, they love stroking their beards and wanking off into the sunset."

Such was the attitude of the center's new thirty-something director. Richard Jonas, now chief surgeon at Children's Cardiac Center in Boston, was one of Mee's early residents in Melbourne; he recalls that his chief made a lot of waves and was oblivious to the politics involved in getting things done. He credits the savvy of D'Arcy Sutherland, a humane presence in a conservative system, with ameliorating the abrasive and alienating effects of the battling boy surgeon.

"Roger was very strong and very righteous about what he was doing," Sutherland himself tells me. "He was succeeding, and a lot of people didn't like him. There he was in his thirties and already at the top of the tree."

Mee furthermore alienated the medical establishment in the state of Victoria and beyond by inviting other centers to send *their* congenital heart cases to him. He knew that if heart surgery on babies and kids remained spread out among half a dozen centers, the way it was when he arrived, the quality of care would suffer badly. Volume, he knew, created good results.

Sutherland asked his colleague Nate Myers, one of the senior pediatric surgeons there at the time, for his recollections, and Myers in turn asked others. Beyond "Roger's excessive smoking, which we all deplored," Myers didn't have problems with Mee, but others did.

"Firstly," Myers wrote to Sutherland, "I must admit that some of the comments I received in reply to my various inquiries surprised me—although others did not. Personally I found Roger to be an excellent colleague and it was never difficult to establish a good rapport with him, and this too was the reaction of some of the surgical staff whose opinions I sought. On the other hand, there were several members of the surgical staff who described him as being ambitious and arrogant, and some even hold stronger views. This was news to me, but at the time of his involvement with the RCH, I was no longer involved in any aspect of administration and I had much else to occupy my time and my mind. One very senior member of the hospital staff (nonsurgical) did make some strong comments regarding Roger's attitudes (and arrogance). Roger's relations with the members of board of management, I think there is little doubt that they were very mindful of the superb job Roger was doing and grateful for the fact that cardiac surgery was again on its feet, as it were. However, the chief of surgery at the time was approached by the board to discuss various matters with Roger, specifically in relation to his persisting demands. The above-mentioned chief of surgery stated that Roger was, in many respects, the most difficult member of the surgical staff—was in fact arrogant in his demands. I think this has to be taken in context, and I have little doubt that personality factors were relevant to such comments."

Sutherland himself saw immediately that Mee "had a feeling for, an understanding of, pediatric surgery. He has an incredible technical ability with his hands, magnificent skills. But Roger had another ability which put him ahead: he had a knowledge of electronics, and his knowledge of physiology was far ahead of anybody else's. He worked well in a team, he was dedicated, and he was great with parents. That's a whole bag of things that make him pretty special."

Within a year Mee had reduced infant mortality from 63 percent to 18 percent and nearly halved general open-heart mortality (from 9.5 percent to 4.8 percent)—figures all hospitals were required to report to the National Heart Foundation, which monitored results. The year after that his infant mortality jumped to 30 percent, largely because the number of cases he handled nearly tripled, to twenty-nine, a figure that would continue to escalate. It took time and continual struggle, but he slowly increased the quality of his center's equipment and of its care to approach the quality at the centers where he'd trained, especially Boston, where he'd been surrounded by an extraordinary team. He hired a good young fellow out of Green Lane named Bill Brawn, who would stay for seven years. He enhanced the quality of perfusion and anesthesiology and postoperative care. (Mee's P.A., Mike Fackelmann, told me that what had impressed him was that Mee could do *anybody's* job in the O.R.—and do it better.) Mee brought his own excellent surgical technique to the operating table, which translated into reduced use of circulatory arrest and reduced cross-clamp time, and in turn to less time in the ICU and improved overall recovery. He had an intuitive sense of the bypass machine and what kind of flow could be run through babies' delicate vessels without destroying them. At the time, his hospital, and others, believed it was safe to leave babies on circ arrest for a full hour; Mee disagreed, attributing postoperative brain damage, seizures, and mortality to the practice. Keeping kids on bypass might make the surgery a little more difficult, but at least it didn't require completely cutting off all circulation.

As word of Mee's results spread, more and more complex cases were referred to him, and fewer and fewer other centers offered pediatric surgery, and finally he was handling all peds cases from Victoria, Tasmania, and Southern and Western Australia. Within five years Mee was doing more than 500 cases a year and reporting a mortality nearly on a par with Boston's. Success brought even more cases: in 1988 he alone performed more than 600 heart surgeries, an astonishing number. That year he opened the country's first pediatric heart transplant center and did its first heart-lung transplant (it was *his* first transplant, too; he'd learned the technique from the movie he took at Stanford, and studied other centers' protocols to create one of his own). In 1991 Mee was able to compare his center's figures with those of Children's in Boston, the historical seat of congenital heart surgery and by reputation the best there was. From 1984 to 1990 Mee's center performed 294 biventricular repairs on neonates, with a 2 percent mortality; during roughly that same period Boston reported an 11.8 percent mortality out of a total of 304 neonatal cases. Mee's results were now the best in the world, by a substantial margin.

The head cardiologist at the Royal Children's urged Mee not to publish his data for fear that people would think he was lying, an imagined deceit that might nonetheless undermine the credibility of the program. Mee published his results anyway. Sutherland recalls, "He did twenty-five simple transpositions, his first, without losing one—nobody believed it." Chuck Fraser, now chief of congenital heart surgery at Texas Children's Hospital, accepted a fellowship under Mee in 1990 after leaving Johns Hopkins—despite being told flat out by other surgeons, he says, when he was still trying to make up his mind about going, "Roger really fabricates this data." Within three weeks of his arrival, Fraser knew not only that Mee was *not* fabricating data but also, and more important, that his was the model to emulate.

When I questioned him about Mee's results, Sutherland, now eighty-eight years old and speaking from his home in Adelaide, said, "Those figures are absolutely watertight honest," adding, "What Roger did and the standards he set were better than anybody else's."

"Roger was better than any of them," recalls Boston pathologist-cardiologist Van Praagh. "He came to meetings with results so much better than ours. His results were so good it was embarrassing."

Mee says one big reason for the Royal Children's success in the 1980s was an idea he had about using a drug called phenoxybenzamine, which was denigrated then—and continues to be dismissed today by some—as old-fashioned. A peripheral vasodilator, it had previously been used unsuccessfully to treat irreversible shock and a particular kind of tumor, but it was known also for its salutary effects on bed wetters. It worked by easing the tension on the body's entire vasculature—which made it a promising drug for use in heart surgery. Mee needed government approval to try it on pediatric cardiac patients, and once he'd gotten that, he found he was right: by dilating the vessels of a neonate who needed to be put on the heart-lung machine, phenoxy enabled the perfusionist to triple the flow of blood to the patient without increasing the pressure, thereby delivering more oxygen to other organs that were typically deprived during bypass—including the brain, the kidneys, the skin—with less damage to the newborn's delicate vasculature.

Little fanfare attended Mee's presentation of his results, superlative though they were. Surgeons can be a skeptical lot, and few, Mee says, believed his data. Most in the field note how easy it is to be selective with results, manipulate them in a favorable manner. Mortality figures are, of course, slippery things and often have scant meaning; some surgeons say they simply don't play that game. There are too many variables involved, and the cases are too varied in their complexity. Mee himself has said that virtually every patient ought to be in his or her own subcategory. But the way credible information circulated then, and always has, was by what in a different context would be termed gossip: word spread along the small, tight grapevine. As fellows and residents worked under Mee, saw his results for themselves, and left for other hospitals, they could confirm what he was reporting. Such firsthand information was more trenchant than anything published in the surgical periodicals, which often appeared many, many months after the fact, at a time when advances were rapid.

Mee began in Melbourne just as prostaglandins were being introduced in all developed countries, transforming the congenital heart world. Where once the lives of the entire O.R. team had revolved around doing emergency surgery—rushing into the hospital in the middle of the night, the surgeons opening the chest urgently and diagnosing the defect on the spot and attempting to palliate it with a shunt or a band around the main pulmonary artery—the discovery, in 1975, that prostaglandin E1 kept the duct open in newborns (along with Castañeda's advances in performing surgery on younger and younger babies) meant that infants could be stabilized, properly diagnosed, and cathed, if necessary, with a surgical plan being carefully thought out before the surgery was even scheduled. Even then, though, surgery on a baby who was hemodynamically stable—that is, relatively healthy—was typically more successful than surgery on a sick baby. In many cases surgery wasn't even possible for the sickest ones. It was futile to shunt a hypoplast, for instance; a baby born without a functioning left ventricle, even if in every other way perfect, would be given to the mother to hold until he or she died, an eventuality that sometimes took days.

The early 1980s would be the beginning of the present era of advances in congenital heart surgery. By then Mee, with Bill Brawn, had lowered neonatal mortality to 2 percent at the Royal Children's, proving that even complex cases could be performed routinely and safely. Others were not far behind. Castañeda in Boston demonstrated through his results that elective neonatal surgery was not merely an option to be reserved for a very few cases; rather, it was, for a range of defects, the advisable course. Prostaglandins provided a huge measure of safety, even as technological advances accelerated further change. Two-dimensional echocardiography dramatically improved diagnostic capabilities. Smaller instruments, miniaturized cannulae, and a better understanding of the mechanics of cardiopulmonary bypass all served to minimize the damage done to infants' blood and lungs, improving mortality and enabling not only neonates but even premature babies—sometimes weighing less than a kilogram—to tolerate open-heart surgery. In Boston Bill Norwood was developing the unimaginable: a palliation for hypoplastic left heart syndrome, a lesion that most surgeons had always assumed would remain unbeatable. In doing so he was solving one of the last riddles of congenital heart disease: how to save a child who was missing the main pumping chamber of his or her heart.

Mee and his fellow Bill Brawn were meanwhile advancing the repair of other previously difficult lesions, such as interrupted aortic arch with a VSD. Building on Norwood's experiments with hypoplasts and adding their own modifications in manipulating the aortic arch, they transformed a two-stage into a one-stage operation, and from these early successes—using the same technique of dissecting out the descending arch so they could pull it up and work with it, thus avoiding unnecessary grafts of foreign tissue or, worse, Gore-Tex—they figured out how to fix still other lesions associated with disfigured arches, including truncus, coarct, hypoplastic arch, and certain kinds of transposition.

Around this time a small group of patients had begun to appear with a *new* problem. They were not babies but rather adolescents, born in the 1960s and 1970s with transposition that had then been repaired with a Mustard or a Senning. Their desperate parents were bringing them in in heart failure because their right ventricles couldn't do the work anymore and were crapping out. What could Mee and Brawn do? These patients had been palliated before the advent of the arterial switch, and now they were young men and women with full lives—friends, lovers, ambitions, work—and they were dying. Extrapolating from the work of Senning and Yacoub, Mee reasoned that if he could squeeze off the pulmonary artery, it might be possible to retrain a left ventricle that had previously pumped blood only to the low-pressure lungs to pump harder. And after a year or more of pumping through the band, the ventricle might be strong enough to take over pumping to the body; if so, he could perform an arterial switch. The first patient he did was a seventeen-year-old who's still alive today. Mee wrote widely about his technique of banding to retrain the left ventricle, then taking down a Mustard to do a switch. He wrote about his modifications of other people's surgeries. He described how to handle intramural coronaries in patients with transposition. And he began performing a very difficult and high-risk procedure called a double switch, which was really two operations in one: a regular switch and then a Senning. With this he reintroduced an operation that was nearly extinct, first because few could do it properly and second because of the success of the arterial switch that had eclipsed it. Here, too, Mee's results were unsurpassed.

Boston continued to be the main training ground for congenital surgeons. By the year 2000, Castañeda would estimate, some forty surgeons who had trained at his center were now the chiefs of congenital heart centers across the country. Ed Bove, a New Yorker who had trained in England, was publishing exceptional results at the University of Michigan, as was Jan Quaegebeur at Columbia Presbyterian in Manhattan. CHOP, the Children's Hospital of Philadelphia, had grown into a major, highly regarded congenital heart center under chief surgeon Tom Spray. And a young Frank Hanley in San Francisco, only recently out of Boston, had begun to advocate the earlier and earlier complete repair of lesions that had always before been repaired over the course of two or three surgeries; and he had the results to support his claims that this approach worked. These surgeons and many more like them were Castañeda's legacy.

Hanley had arrived at UCSF in 1992, ending that university's search for a marquee surgeon. The original hope had been to lure Roger Mee there, and Mee had considered it. He had taken his center at the Royal Children's as far as it could go for now, had developed an extraordinary system that ran smoothly. He was forty-eight years old—did he want to ride this one out, or was it time for him to start again elsewhere, to re-create a center as he had done in Melbourne? He was a builder by nature, so he and Helen decided to try one more move. He declined the offer at UCSF. When word got out that Roger Mee was shopping around for a center, thirty more offers came in. Two centers had put together promising packages. Mee called Castañeda, then in his last year at Boston, to ask his advice, and Castañeda recommended Cleveland.

On October 14, 1992, Mee wrote to the clinic's chief cardiac surgeon. He began, "Dear Dr. Cosgrove, As agreed, I am setting down my thoughts on paper concerning the development of a congenital cardiac surgery service at the Cleveland Clinic," then went on to list his many requirements for a separate congenital unit, including staffing for ICU, O.R., administration, perfusion, anesthesiology, and nursing, necessary equipment, and protocol. He concluded his introduction by declaring, "I am happy at this point to indicate my own personal belief; that in general terms, the delivery of excellence and the submission to political compromises are probably mutually exclusive."

Mee's debut in Cleveland was a PDA ligation, the operation first performed by Robert Gross in 1938 to usher in the age of elective heart surgery. The date was August 11, 1993. By the time I came to observe at the clinic in that grim fall of 2000, when the ICU was taking its pounding, when Frank Moga told me, "This is what you shouldn't see, this is the horror," when Drew H. got his God-sent heart while his near equal in weight, Ashley Hohman, struggled to hang on, Mee's center had performed more than 3,000 operations and was recording its most successful year ever: 575 operations between him and Jonathan Drummond-Webb, with Mee himself doing three quarters of them, and just five deaths, an overall mortality of 0.9 percent. Mee didn't know of any better results in the world, certainly not at a center like this one, which attracted FUBARs no other hospital would touch. The results were so good, in fact, that a surprised Mee confided to me in a serious tone, "I think we just got lucky."

These results were the very thing that made this specialty different from almost every other, because in the end they stood for babies and young children and whether they lived or died. Mortality at Mee's center might rise or fall a point or so from year to year (the previous year overall mortality had been relatively high, 2.95 percent, but the general average was about 1.6 percent), as it did at all the best places, such as Boston and Michigan, which also regularly recorded overall mortality of less than 2 percent while performing large volumes of the highest-risk surgeries. But few centers in the country could boast such outstanding results; the national average was closer to 5 percent, according to data provided by the American Heart Association. And some—often the centers that handled fewer congenital cases—recorded much higher mortality; for complex cases such as switches and Norwoods, the rate could be 50 percent or more. After seeing how terribly wrong it'd all gone for Drew H. until he got here, I couldn't help but wonder why these subpar places didn't refer out their toughest cases. These were babies and children. I'd come here to watch people at work in an excellent heart center. But I wasn't a medical professional; I was a parent, and I identified with these parents—I couldn't help but put myself in their position. Especially as it became increasingly clear that, as one of the cardiologists here would say to me, "The parents don't know."

7. The Physician's Assistant

"There's nothing worse in the world than losing your child. Nothing. Your *children*. There is nothing worse than that." Mike Fackelmann, himself the father of two teenagers, has been working as a nurse for eighteen years now, almost all of that time as an O.R. cardiac nurse, and since 1993, not long after Roger Mee arrived at the Cleveland Clinic, as Mee's P.A., or physician's assistant. The job is hard, he says, but he likes it—it's "way cooler" than being an anesthesiology nurse, and it beats his first job managing an assembly line at the Ford Motor Company's Brookpark plant. The work was harder before Roger came, when Mike assisted the surgeon who did the clinic's congenital cases. He says the job fell to him in part because he was the only person who got along with the peds surgeon. Mike insists, "He was a great guy," but everyone else I spoke with described the previous surgeon as "difficult"—and given that *most* surgeons are considered difficult by default, being singled out by colleagues for that quality takes some doing. As far as his results went, no one had any particular praise or criticism to offer. "He was an average surgeon," said a cardiologist who was there at the time. "His numbers would compare with most others in the country." June Graney, the cardiac nurse, handed over her own baby to the O.R. staff so this surgeon could ligate her patent ductus arteriosus (PDA). And Mike Fackelmann got along with him. Sometime during that period Mike was taken aback by a casual remark made by Floyd Loop, then chief cardiac surgeon and soon to be the first heart surgeon turned CEO of a major medical institution. He was the head guy, the emperor of a world-class cardiac surgery unit, a talented cutter and a powerful personality. Mike stepped into Loop's O.R. one day to take a vein—for the CABG procedure a technician must remove a large vein from the patient's leg to graft onto the blocked coronary, a process that's referred to as "taking vein"—and Loop's P.A., introducing him, said, "Dr. Loop, this is Mike Fackelmann."

Emperor Loop said, "I know you—you work with kids."

Mike said, "Yeah, that's right."

And Loop said, "Thanks."

Only words from God would have been more meaningful to Mike then ("That guy does probably the best distal coronary anastomosis of anybody in the *world*," he says). Loop did not elaborate on his thanks, never specified whether he meant "Thanks for putting up with the surgeon" or "Thanks for doing peds," the most depressing of the cardiac-surgery subspecialties.

So Mike was the main peds assist, but the work was dispiriting, because mortality ran about 10 percent out of the approximately 180 cases a year handled by the center (and since the majority of that total—a hundred or so cases—were adult congenital patients, who tended to come through surgery fine, the mortality occurred primarily in infants, intensifying the impact). One of Mike's worst experiences was assisting on the daughter of a friend of his. She had a ventricular septal defect (VSD), and it wasn't patched very well; a residual defect remained after the operation, so the girl had a hard recovery and wouldn't ever be as healthy as she might have been had she received better treatment. To this day, more than ten years later, Fackelmann cringes every time he sees her. He didn't do the surgery himself, of course, but he feels responsible nevertheless: he'd promised his friend that the team would do an excellent job, but they didn't. In peds heart surgery, Mike says now, "nothing is 'good enough.' It's got to be perfect. There's no compromise."

When a young patient dies, it's always hard. George Thomas, the perfusionist who did most of the peds pump cases back then, says he hated the work because of the mortality—"I'm a father," he tells me by way of explaining how awful it was. When Mike worked for Mee's predecessor, he would go home and feel brutalized for days, thinking, he says, "We killed that kid." Mike says now, "There were guys out there getting great results, but we didn't know that; this was all we knew." And it was all that most here knew of congenital heart disease: all-day operations, hard recoveries, 10 to 20 percent mortality. The nurse covering such a service could offer only this as advice: "Don't get born with heart disease."

Mike Fackelmann, with his small, trimmed mustache, his blond hair receding over his oval brow, is a good-looking guy of forty-five who walks with a subtle strut: even his footsteps are definitive. Or maybe a strut is inevitable for someone with Emily Post posture, because if anything, Fackelmann is a mass of insecurity outside the O.R. He is amazed when someone compliments his work; he brushes it off. After I stopped going to the clinic on a daily basis, I told Mike what Mee had said about him: "It's ironic," Mee had told me. "Mike will never be a surgeon, and he will never operate on people. And yet he's ten times better than the fellows, and they're going out and operating on people."

When I repeated this to Mike, he was flabbergasted: "You're kidding me. He really said that? Are you bullshittin' me?"

Mike is the kind of P.A. who gets clandestine job offers via e-mail from other surgeons who have come to observe Roger Mee. (There's such a steady flow of visiting surgeons that if I really needed to watch an operation, I'd have to get to the O.R. while Mike and the fellow were opening to ensure I got a good spot.) It's easy for visitors—even me—to see what an asset Fackelmann is in the O.R.

He's great to watch and compelling to listen to. He doesn't talk in the O.R. once the patient's chest has been opened unless something needs to be said ("Stay away from that," for example), so when he *did* speak, I made sure to pay attention. He'll even tell Mee to back off if he sees some danger, and the heads-up is always justified, whether or not Mee chooses to listen to it, because Mike can do what Mee can't: keep his eye on the monitor. He can look up when Mee is tugging on a critical vessel and watch pressures drop, or notice segment changes in the EKG, meaning that the heart isn't liking what Mee is doing, and then he can warn Roger. He's there to assist. He's there to watch Roger's ass, because Roger has to be able to concentrate on the repair, thinking three, five, ten steps ahead, and he can do that best when he trusts that someone is looking out for him. In fact, Mee doesn't like to schedule particularly difficult cases when Fackelmann isn't on.

I had no idea how valuable Fackelmann could be until one afternoon when I stepped into Jonathan Drummond-Webb's O.R. Roger was away, and Jonathan had been dealing with a lot of rough cases—serious problems like a single-ventricle heart with obstructed veins, a cluster of defects associated by some estimates with a 95 percent mortality, and Norwoods with tiny ascending aortas—but this should've been an easy one. The kid—an eleven-week-old baby with tetralogy of Fallot complicated by pulmonary atresia (no pulmonary valve opening) and a tiny main pulmonary artery—was very sick, but the intended palliation, a central shunt in a closed-heart operation, was not a high-risk procedure. It wasn't even a pump case, no cardiopulmonary bypass. I was about ready to leave when Jonathan said, "Fuck . . . We're in fucking trouble." That's when your stomach turns over, because the phrase "fucking trouble" simply isn't used unless there *is*. *Trouble* can mean one of two things: death or near-death resulting in severe brain damage. Anything short of this is *expected* in pediatric heart surgery and therefore not considered trouble. Jonathan was in his hood and headlight and loupes, a scrub nurse at his right, a resident at his left assisting, Charlie across from him, and Mike next to Charlie—five in all scrubbed in, with Julie Tome on anesthesiology. The room was quieter than normal because there wasn't any hum from the refrigerating unit of the bypass machine—just the normal beeps from the monitor, an occasional bovie zap, and some loud banter from Jonathan.

Till that moment all had been calm, a normal procedure, afternoon sunlight coming in through the window. Jonathan had evidently decided to do this case off pump to avoid the ill effects of bypass, which always hit the body hard. He'd poked around the beating heart to scrutinize the situation, then ligated the main pulmonary artery, which wasn't doing much of anything (the baby's lungs were being fed by collateral arteries that had grown from aortic vessels into pulmonary arteries—referred to as MAPCAs, or collaterals). He then asked for a U-shaped clamp called a Castañeda and clamped off a small part of the aorta, a maneuver called a sidebite. This way he could maintain the flow through the aorta while cutting off blood to the small section that he intended to attach to the distal pulmonary artery. The only hitch was that the part of the aorta that he needed to sidebite and sew to the pulmonary artery was at the back, meaning that he had to rotate the aorta up toward him. The danger here—the reason another surgeon might have chosen bypass despite its disadvantages—was that it was easy to kink the coronary arteries and thus cut off the blood supply to the heart, a situation the heart could not tolerate for long. Also, tying off the main pulmonary artery, clamping part of the aorta, and twisting it made it harder for blood to leave the heart; if the organ got too full, it could begin to resemble a balloon that had been blown up and then deflated, becoming enlarged and flaccid and unable to contract. If the baby had been on pump, it would've been a simple matter to drain the heart and avoid the problem altogether.

So after Jonathan sidebit the aorta and rotated it to the position he needed it to be in so he could do the sewing, he'd paused to check what the effects of his manipulations were. "Julie, I'm clamping the aorta now," he said. "How are our pressures?"

Julie looked at the pressures on the monitor and said, "Pressures are fine." Scanning all the info on the screen, she added, "Everything looks fine."

Jonathan said "OK" and cut into the aorta.

No sooner had he made the no-going-back incision than Julie said, "We're getting some irregular rhythms."

"Fuck," Jonathan said. "That's why I asked you!" He looked up quickly at the monitor, saw the S-T segment changes, and realized he was kinking a coronary, cutting off blood to the heart. "We're in fucking trouble."

Fackelmann said, "Just sew, Jonathan, just keep going." His voice was quiet, even casual. Jonathan had to sew a small circle to connect the two arteries back to side. He remained furious, but he continued to sew rhythmically. The heartbeat grew irregular, and the ventricles failed to expel all their blood; the kinked coronary had reduced the heart's ability to contract, and it was now ballooning out. Fackelmann slid his hand into the chest, softly rubbed the ventricle with his fingers to feel it, then pushed on it, just so, in a gentle rhythm. He had sensed that the heart was too full, knew it needed help pushing blood into the aorta and head vessels and into the coronaries. He watched the pressure monitor and used his sense of touch to gauge how firmly to push in this open-chest message.

Still seething, perhaps a little panicky now, Jonathan asked, "Is Emad in the room?" He wanted the head anesthesiologist in there, was in effect telling Julie she didn't know what she was doing. "Do you have blood in the room?"

"Yes, I do, Jonathan."

He continued to sew and asked, "Do you have help?"—meaning, *Where's Emad?*

"I have help."

"Don't argue with me!" he said loudly. "Get Emad in here."

"He's coming," she said evenly.

"Come on, Jonathan, just keep sewing," Mike repeated.

"I want you to have twenty cc's of blood ready."

"I have blood ready," said Julie.

"I'm not trying to be difficult. I want you to put it in when I unclamp the aorta." He was almost finished with the sewing. "Don't give her blood until I tell you to. Have you given bicarb?"

"Bicarb is in. Everything is in," Julie assured him.

Fackelmann said, "It's looking better. We're OK."

"We're *not* OK," Jonathan insisted. "There's air in the coronaries." Another implied indictment of Julie's skills.

"I don't know how air could have gotten in the coronaries," Julie said.

"Do you want to come here and have a look?" Jonathan asked. "I can see there's air in the coronaries. Just pull back on your lines."

"I already did that."

Air in the coronaries is yet another kind of problem. It's not the air itself that's bad; what's bad is that it prevents blood from getting to the heart muscle. Air must be pushed through the coronaries, across the capillaries, and into the venous system before it becomes benign. Having air in the coronaries is like having little rocks in there. It could have come from one of Julie's lines, or it could have gotten sucked into the right ventricle when Jonathan transected the pulmonary artery, then pushed through the VSD and into the left ventricle and the aorta. However it had happened, the heart was very full, noticeably bloated, so Mike continued to administer CPR with his left hand—too much pressure could damage the muscle, but too little wouldn't help it eject enough blood—as he used the sucker in his right to keep Jonathan's field clear, all the while watching the monitor.

"It's looking good," Mike said matter-of-factly.

Jonathan tied the stitch off and removed the clamp. The heart picked up its rhythm, and the oxygen saturations jumped from 40 into the 70s.

Mike looked at Jonathan and said, "I guess your shunt is working."

Jonathan stopped now. He stretched out his arms so his hands were on the edge of the table, leaning hard on it, and hunched over the patient, head bowed. "I saw this kid leaving in a *box*," he said. More quietly, to himself, he murmured, "I should have done this on bypass. Damn." He shook his head.

Fackelmann said, "It's always like this with these."

They began to close. Jonathan asked, "Should I put a PD cath in? This kid is really sick."

"Might as well do it now," Fackelmann answered.

Now it was as though nothing had ever gone wrong. Jonathan was closing the chest and talking and joking, telling everyone about a commercial he'd seen for Alaskan salmon from John West Foods, in which an apparently crazed man fought a bear for a freshly caught fish. It was hilarious, he said, and he hooted.

He turned to Mohamed, the resident in the "monster boy" position, and sent him away, saying, "You were on call last night"—meaning, *I'll close the chest,* a job usually left to Fackelmann and the resident on the case. Mike urged Jonathan to apologize to Julie. "Come on," he said quietly, "we weren't nice to her." *We*. Jonathan said, "Julie, I apologize." Emad had come to the O.R., seen that everything was under control, and left again. At that moment Julie had dipped behind the cage to check lines and was out of view. Jonathan, sewing, said, "Julie? Is Julie in the room?"

"No," said Julie's voice.

"Well, I was going to apologize, but I guess I can't if she's not in the room."

Later I asked Mohamed about the case. This confident, dark-skinned Indian, who was concentrating on left ventricular assist devices and end-stage heart disease in adults, had a voice like James Earl Jones's. "That was really nice," he said when I mentioned what Fackelmann had done that day in the O.R. "That's not easy to do. It showed finesse, experienced hands." Mohamed was a young surgeon, and he was impressed.

Fackelmann himself told me afterward, "They behave like little kids, they're just like little kids. . . . [Jonathan] can't see the forest for the trees, isn't that it? He shouldn't have been worried about what Julie was doing—at that point it didn't matter. What mattered was getting the clamp off the aorta." When I asked him to elaborate on what he'd done in there, how he'd known what to do, he told me it was all instinct—"Every little thing you do," he explained. "You see the pressure going down, you see the heart slowing down, you see it changing color, you don't even have to think that you've got to do massage to get the pressure up; you just automatically do it. You've got a resident who's standing there not really knowing *what* to do. Because residents don't. They're *scared*. I don't know. You don't even have to think about it. You know what to do."

A mood of both excitement and trepidation had attended the arrival of Roger Mee in 1993. On the one hand, there was the predictable buzz among the peds cardiac staff, who'd been thrilled to learn that one of the best peds heart surgeons in the world was joining the clinic, but on the other, they'd also heard rumors that he was bringing his own team with him and that he didn't use nurse clinicians, which made their own job security an issue. If the rumors were true, Mee wouldn't allow nurses like Mike to assist in his cases, wouldn't even let them anywhere *near* his patients.

In the end Mee did bring in his longtime secretary and coordinator, two perfusionists, and some of his own equipment, and also recruited an American surgeon, Chuck Fraser, who had been a fellow under him in Melbourne. Like a celebrity chef landing in an unfamiliar kitchen, he knew that he'd have to come prepared with the essentials, expecting to find in his new workplace nothing but the basics. Mee was well aware that most hospitals that absorbed peds into bigger adult programs tended to treat babies like miniature adults, when in fact they were unique and hugely complex, terrifically fragile systems varying greatly by weight, especially when it came to anesthesia and perfusion—in part because their surface-area-to-body-mass ratio differed so significantly from that of adults. Mee had some inevitable we-do-it-this-way-well-I-do-it-*this*-way clashes with the clinic staff (though nothing along the lines of his recurrent nightmare, in which he was offered rope instead of Prolene and told it was what the old surgeon had used). Some disputes were minor, as when Mee explained that he rather fancied cigarettes and didn't intend to give them up, and said he hoped no one had a problem with that. (The clinic installed a special machine in his office to suck up the smoke.) Other issues were more contentious, such as Mee's insistence that new O.R.s be built in another part of the hospital so congenital heart surgery would be completely separate from adult heart surgery, and the notion of splitting off pediatric cardiac ICU care from adult cardiac ICU care, which Mee was adamant about and to which his boss was equally adamantly opposed.

Delos "Toby" Cosgrove, the chairman of thoracic and cardiovascular surgery at the Cleveland Clinic, was the man responsible for hiring Roger Mee. A tall, athletic-looking doc then in his mid-fifties, Cosgrove had trained in Boston with Castañeda and helped write a book on congenital heart surgery; he was acknowledged as being among the top heart surgeons in the world, noted among other things for his innovations in mitral-valve surgeries (of which he did between four hundred and five hundred a year) and widely respected for having put together the world's best cardiac surgery unit. He was, in short, a surgeon in a powerful position. In person he was the model of the confident, top-notch heart specialist, but in the O.R., according to one tech, he was a "bull in a china shop." He had a strong personality and, I'd wager, a colossal ego (they all did, and most admitted it without apology).

Recognizing when he took over the department in 1990, after Floyd Loop moved into the clinic CEO's seat, that the congenital segment of the service was lacking, Cosgrove also knew exactly how it had to change. "Either I was going to do it world-class, or I wasn't going to do it at all," he says now. At just this time, word got out that Roger Mee was considering offers. Cosgrove approached him on behalf of the clinic, and Mee ultimately said yes. Before he arrived, Mee reiterated in writing to Cosgrove his "desire to set up a congenital heart program rather than just [do] paediatric cardiac surgery."

Shortly after he got to Cleveland, Mee realized that a separate congenital center was not part of Cosgrove's plan. Cosgrove wanted all heart surgery to remain within his purview and did not believe that excellence in congenital required separate services. Mee disagreed and fought his chief. When it began to appear that he wasn't going to get his way, he decided to return to Australia, and called Helen there to say, Stop packing, I'm coming home. "They just want a trophy on their wall," he told her angrily. Loop, traveling in the Middle East at the time, got wind of the strife and telephoned Mee to intercede, promising him what he wanted and persuading him to stay.

Mee chuckles now and says of his tête-à-tête with Cosgrove, "You put two bulls in a ring, they're going to fight."

Cosgrove, for his part, said in an interview, "Roger is a master congenital heart surgeon. His results are second to no one's." Reserved about his personal feelings, he would tell me only that Mee had decided that adult and congenital could not be combined, "so that little empire grew up over there." End of story. No one likes to lose, and heart surgeons, I think, like it less than most.

Fackelmann, one of the serfs gazing up at the castle, says he doesn't "know how that relationship disintegrated."

Mee had left the Royal Children's with an option to return there if Cleveland didn't work out for him—he was well aware of medical politics from his struggles in Melbourne and had told Cosgrove outright that in his view, political compromises were antithetical to the provision of excellence. As the outsider being invited in, he had a couple of things working in his favor. First, his results *were* second to no one's; if the clinic wanted the best, it would have to meet his demands, or he'd take his toys and go home. Second, the delivery of great cardiac care has the helpful side effect of generating a lot of revenue. While most children's hospitals perpetually struggle with red ink and bottom-line issues, cardiac surgery is often a cash cow. The clinic was intent on creating a first-rate children's facility, a goal underscored by the fact that pediatrics had been elevated to its own division (rather than kept as a department within a larger division) and by the designation of part of the huge clinic complex as a children's hospital (it would never be a standalone unit). A world-class congenital heart surgery program could be a financial linchpin and shining emblem, inasmuch as it would generate both the attention that always surrounds a great heart surgeon and sufficient income to make up for all the money that would be lost in infectious diseases, endocrinology, and other non-procedure-oriented specialties. According to Richard Jonas, the head of congenital surgery at Children's in Boston, children's hospitals ride such a precarious razor's edge financially that "they almost *have* to offer heart surgery to survive."

And if not cardiac surgery, then the children's hospital must develop some *other* core program that will be equally procedure-oriented and technically sophisticated. In big pediatric hospitals—say, those with more than two hundred beds—a single such program will represent a smaller percentage of the hospital's revenue. But at a small or medium-size hospital such as the clinic's (with just over ninety beds), that program will be a meaningful enough contributor to revenues (it "can equal well over half of the hospital's operating margin," according to the clinic) to suggest, *No heart center, no children's hospital.*

· · ·

When Mee first arrived, and the previous surgeon had been dispatched, Mike Fackelmann didn't have a regular place in the congenital lineup. He could work in the O.R., but he had no contact with patients. "He wouldn't even let me put a Foley in," Fackelmann remembers of his new boss. But the main scrub nurse, Debbie Seim, knew that Mike was an asset. Debbie had traveled to Melbourne to observe Mee's team so he'd have a scrub nurse in Cleveland familiar with his system. Not long after Mee started at the clinic, word came up to the O.R. that a baby had been born at an outlying hospital with critical aortic stenosis and was being transferred *now*. This was going to be a hairy one, Deb knew, so she paged Fackelmann, who was on call that weekend. Thinking he had a long-distance pager, Mike had driven more than a hundred miles south to spend the day at the Mid-Ohio Racetrack, a big auto-racing center in the middle of the state. It wasn't until he headed back to his car to grab another beer that he heard his name echoing out of the track's loudspeakers. He called home, his wife told him he'd been paged, and he hopped into his car and was at the clinic before the baby got there.

The baby was "actively dying" by the time they hustled him into the O.R. and began an urgent procedure to give him CPR and at the same time crack his chest. (This procedure looks like a free fall, with the nurses squirting Betadine on the hands of the surgeons while they're working on the chest, alternately pumping and cutting.) Mee was on one side, a cardiac resident was on the other, and Mike had scrubbed in to assist. The resident soon proved to be in over his head—was utterly lost—and Mee simply said to Fackelmann and the resident, "You two switch places." After they got the chest open and got the kid on pump, the rest went smoothly: Mee opened the aortic valve with two incisions and closed the vessel up. The patient recovered well.

"After that," Fackelmann says, "he let me be in the O.R."

Fackelmann might have made a great surgeon if he'd pursued medicine after college, but his working-class background didn't offer him that option. (It did, however, provide him with good advice for buying a car: always check the engine manifold, where there's a coded date noting when it was put together; never buy a car whose engine was put together on Friday—the assembly-line workers get slack just before the weekend, Fackelmann explains.) By chance, though, he found work that seemed perfectly suited to his desires and temperament. When I ask him if he likes his work, he looks at me as if I were crazy. "*Yeah*, I *like* what I *do*," he says. "Because it's fine work. It's *fine*. It's elegant. It's precise."

In addition to his work as a nurse and P.A., or first assistant, as the job is alternately called, Mike attends conferences to lobby for better pay and benefits for himself and those of his colleagues who have likewise earned certified registered nurse first assistant (CRNFA) status. He loves going to these gatherings, in part because it's good for his fragile ego: "Wherever I go—if I go to Dallas, say—people know who I work for," he says. "I feel fortunate to work for Roger, for someone like that."

Fackelmann is the noble foot soldier in General Mee's daily battles against defective hearts, but that's all he is. "We're not friends, we're not buddies," Fackelmann says, and yet he adores having Roger's attention, like the kid brother of a star athlete; they don't converse often, but Fackelmann beams whenever they have a nonwork conversation. And he can be easily hurt. He once called Mee at home about what he said was some hospital business (actually, it regarded a paycheck prank by Mike's colleagues, who knew it would drive him crazy), and Mee told him not to call him at home again. This was better than saying he'd reach through the phone and rip Mike's fucking trachea out if he ever bothered him at home again, as another heart surgeon, taken away from some Sunday gardening by Mike's call, did once. But Fackelmann didn't think Mee's response was right either.

On another occasion Fackelmann, wearing his blue scrubs, with paper booties over his white bucks, strutted up to Debbie Gilchrist's desk outside Mee's office to make sure she and everyone else remembered that he'd be out of town at the end of the week and through the weekend, lobbying in Columbus. Debbie was howling because Jonathan Drummond-Webb's staff-meeting-cum-farewell was accidentally being broadcast in the A building cafeteria. Deb was cracking up, but Mike looked unsettled.

"I wasn't invited," he said.

Deb stopped laughing and gave him a *You're kidding, right?* look. Then she said, "It's a staff meeting. You never go to those."

"Yeah, but it's a farewell. How come I wasn't invited?" He was serious, and a little miffed. Then he asked, "Did Jonathan wear his extra-tight pants?"

Deb laughed again and said, "And no underwear!" Then she reminded Mike that he was supposed to interview a woman who'd applied for a second physician's assistant position in the O.R.

"Who is this person?" he asked. "An OB-GYN nurse? She's not qualified to do this. Why is she applying? This is ridiculous."

"She's waiting for you up front," Deb said. As Fackelmann turned to leave, shaking his head, she shouted out, "In stirrups!"

While he's known to be "difficult" and overly sensitive, Mike can also be edgy with Mee, albeit in an overly sensitive manner. If Mee happens to compliment him in the O.R., for instance, Fackelmann may reply, as he did in one instance, "You don't have to kiss my ass; I'm not here to kiss *your* ass." To me he's claimed, "I just want to be respected for what I do." Fackelmann recalls one time in the O.R. when Mee was talking about the logistical problems he'd had on a recent trip: his passport had not been in order, evidently, so he'd had difficulty entering the country, even after specifying who had invited him and whose child he was there to operate on. Fackelmann laughs at the memory, shakes his head, and says, "If your passport's not in order, it's not in order. He puts on his pants same as we do. Technically great, he's really smart, but he's not God."

Fackelmann is an impressive guy, with a grace and efficiency of movement that make him beautiful to watch in the O.R., and an intelligence and articulateness that make him compelling to listen to outside it. Just as Frank Moga articulated for me the peculiar and intense ethos of the peds heart surgeon, Fackelmann proved a similarly effective guide through the world and language of the O.R. At first he refused to talk to me and recommended that others do the same, but when I didn't go away, he became direct and thoughtful and gave me information that simply wasn't available to me otherwise. For example, when I asked why the operations Mee performed were so difficult, he said, "Try to sew two Kleenexes together using needle holders and seven-oh Prolene, and tie it off without lifting the tissues, because if you do, you've ripped the stitch out of the aorta."

Fackelmann is direct in confronting errors or sloppy work, able to reduce things to black and white (as he has to do every day in the O.R.), in a way even other surgeons don't. Take the surgical error made by a general surgeon operating at a clinic-affiliated hospital.* The case was an infant undergoing cardiac surgery, an intended PDA ligation. Rather than ligating the patent ductus arteriosis—the short vessel running between the aorta and the pulmonary artery, which is supposed to close up after birth—the surgeon ligated, or tied off, the baby's left pulmonary artery. The child, of course, did not recover but instead grew very sick as all blood to her left lung was cut off and the wide-open PDA flooded her right lung. The mistake was discovered, Mee did the repair, and the baby spent weeks in the hospital recuperating from the damage.

When I asked Mohamed whether the surgeon's mistake had been serious, or outrageous, he said no, and I believed him. I knew what it looked like in there, with all those vessels knotted together in that tiny compact space, and knew, too, that the surgeon had only a limited view. "It's one of the risks for that surgery," Mohamed explained. "It's not like it hasn't happened before. That would be bad, to be the first. It would be named after you: 'the Mohamed tie.'" He laughed. "That would be embarrassing." Surgeons

---

*Errors happen all the time in hospitals: surgeons operate on the wrong side of the patient's brain, the wrong leg, the wrong *patient*, sometimes, or forget to pick up their tools when they're finished (in one egregious case a surgeon left a thirteen-inch-long retractor inside a man's belly). They happen often enough, in fact, that the winter after I was an observer at the Cleveland Clinic, a Joint Commission on Accreditation of Healthcare Organizations would issue its second alert in three years, in hopes of eliminating or at least reducing the number of surgical mistakes. In his story on the subject, Lawrence K. Altman, a doctor turned *New York Times* reporter and medical columnist, wrote that this alert "reflects an unsettling fact that doctors and patients may prefer not to think about—that serious mistakes can happen even in the best of hospitals" (*New York Times*, December 11, 2001).

are almost invariably forgiving of one another, perhaps because they're aware that what goes around comes around (though not Mee, he's never forgiving, and when the surgeon who'd made that error came into the O.R. while he was undoing the damage—the manly thing to do, certainly, part of his penance; he surely felt awful about what he'd done—Mee, in the middle of the difficult procedure, was brusque. In any event, Mee and the surgeon had spoken civilly about the case already, he liked and respected this surgeon, and he knew how important it was for general and cardiac services to get along well.

Fackelmann's anger, by contrast, was directed wholly at the surgeon. Errors resulted from carelessness, and carelessness was avoidable. When I mentioned an unintentional error by a surgeon I'd read about, his response was immediate: "What an *asshole*," he said. Fackelmann was angry and incredulous. You didn't touch *anything* without knowing what it was. In his mind, errors were a choice.

I happened to be observing in the O.R. one day when something surreal and—because ultimately no harm would come of it—hilarious occurred. Mee was in the middle of a procedure, his tools down into the hole of the chest and at work on a beating heart, when he said some words that made me look up, they were so odd: "Is somebody *eating* in here?" he asked. At first I thought he must be joking, but there to his left, on the sterile green towels draping the patient, lay a chunk of graham cracker the size of a quarter and some crumbs, immediately in front of the resident assisting in the "monster boy" position. It was a confusing image because it seemed so wrong and strange, a scrap of food there in the O.R. The resident shook his head. Mee lifted the graham cracker with forceps and examined it, and though his face was almost completely covered with gear, his disgust was palpable, an energy radiating off him. He looked at the resident once more over the top of his loupes, then chucked it all on the floor behind him, the forceps bouncing with a *clink, clink*.

Fackelmann spoke up: "*It fell out of your mask*." The resident shook his head again. "*Yeah it did, it fell out of your mask*," Mike said. "*You were eating in the lounge with your mask still tied around your neck, and it got caught there, didn't it?*"

The resident shook his head and said, "No, it wasn't me."

Fackelmann had been in the O.R. the whole time, but he was exactly right about what had happened. I knew this because the resident had been in the lounge, with his mask tied around his neck and hanging down like a bib, talking to me and eating graham crackers immediately before he scrubbed in.

Fackelmann is a maniac for order at home, too. The obsessive precision he brings to his work can't easily be turned off. "My wife and my kids think I'm crazy," he admits. "My kids ignore it; they laugh. ... I don't care about my refrigerator, I care about my garage. I don't care about my wife's car, I care about mine." When I ask him about his dresser drawers, which I've heard he's fanatical about, his response is "I don't want anybody touching *anything*, because it's the way I want it, ya know?"

Fackelmann, like Mee, has a recurring nightmare, in his case one that's descriptive of the sensibility of the fanatical and sensitive physician's assistant. I asked many in the field if they had recurring work-related dreams, but Mee and Fackelmann were the only ones who said they did, and I wondered if it attested to a similar type of personality and a parallel level of intensity in their work. Fackelmann nodded and smiled when I asked, then looked away.

"What is it?"

"I'm taking my own vein," he said.

"What?"

"I'm taking my own vein—I'm assisting in my own operation."

If Mee's dream is about quality control, Fackelmann's runs along the same lines. He says he knows why he has it: he doesn't trust anyone. He wants everything done perfectly, and he can meet that standard, he does it every day, but other people don't. How could a guy tie something off if he wasn't sure what it was, a critical vessel in the very center of a baby girl's body? He didn't understand it, and he couldn't forgive it, but that was the way it was. Given the option, he'd take his own vein, and he'd help sew it into his own heart if he could.

• • •

Makoto Ando is opening, and Fackelmann says it once and twice
and then again: "You're gonna get into it." He watches Mac's every
move, a sucker in one hand and a pickup in the other. The eighteen-
month-old child on the table, chest spread open by steel chest
retractors, is Patrick Cundiff. His heart is dark red and gristly
looking from two previous operations. Opening is a precarious
business when there are so many adhesions. "Come at it from this
way," Fackelmann says. "You're gonna get into it." He insists that
Mac needs to make a cut with scissors from the opposite direction,
and Mac does what he says. They have the pericardium partly open
now, but the procedure is still risky: the pericardium is glued to the
right atrium.

Emad Mossad, the head peds anesthesiologist, asks, "Can we
put the blood away?" (In any chest reopening, the surgical resident
or fellow will ask "Is blood in the room?" before putting saw to
bone, in the event of a catastrophic cut.)

Fackelmann's still on the lookout for complications—this
heart's a mess. He asks Emad, "How long can we keep it out?"

"A half hour."

Fackelmann, watching Mac's moves, says, "Put it away. Can
you give him some volume if we get into the atrium?"

Mossad nods once and says, "Yes."

Fackelmann, having anticipated the main potential problem and
planned how to solve it, tells Mac, "Take your time." And again,
"Take your time." Mac shakes his head and mumbles. This is a difficult one: he's one thin layer of tissue away from disaster. They give
up on that side and look for another way to free the heart without
ripping it open. Fackelmann says, "Call Roger and tell him fifteen
minutes."

Patrick Cundiff has been referred from another hospital after two
surgeries, a central shunt and a B-T shunt, to palliate a defect called
double-outlet right ventricle, in which both the pulmonary artery
and the aorta rise out of the right ventricle, and the right ventricle

thus pumps blood through both outlets, while the left ventricle, with no outlets, shoots blood through a VSD into the right one. This toddler has a small VSD, about the size of a pencil eraser, which contributes to the flow and the mixing, but not enough; what's more, one of the shunts appears to have blocked off, so the baby is persistently hypoxemic, meaning that his blood doesn't carry enough oxygen. His oxygen saturations have been slowly but steadily dropping, and little Patrick hasn't been growing as he should: he weighs just seven and a half kilograms, or about sixteen and a half pounds, three quarters of what he ought to weigh at his age. The referring surgeon believes that the only treatment left is to turn his heart into a single-ventricle pump. Maintaining the two ventricles *and* fixing the problem would seem to be an impossible task now.

"We discussed [the surgeon's] findings at our catheterization conference this morning," the referring doctor wrote to Mee, "and the consensus was to proceed to a Glenn or a hemi-Fontan operation and atrial septectomy because of the complexity of a Senning and VSD closure plus either an LV-to-PA conduit or subpulmonic resection.

"Before committing this child to a single ventricular repair, I wished to seek your opinion on this somewhat complex anatomy."

Committing this patient to a single ventricular repair was a serious and permanent decision. It would mean that the heart would have only one ventricle and thus would be a compromised heart, one that must pump blood to both the body and the lungs; and that, in turn, would mean a forever compromised physical life for this child. Sometimes, in such cases, the work proves to be too much for the heart, and the Fontan fails, resulting in a protracted and particularly awful death for the child and the whole family. Mee notes that sometimes a biventricular repair is inappropriate, that an average Fontan is better than a bad biventricular repair. But here if both ventricles could somehow be preserved, Patrick might have a chance of living a close-to-normal life. Nothing less is at stake here.

"A biventricular repair," says Marc Harrison, one of the ICU docs who would be taking care of Patrick after his operation, "means no more surgeries, normal exercise tolerance. With a univentricular repair, you have a life of certain cardiovascular disease. And we don't know how long it'll last. He certainly won't live to be eighty. But will he even live to be forty? Twenty? We don't really know. With a biventricular repair, after recovery, eventually we'll say, 'Come see us every six months or a year, but go out and live your life.' It's a big deal."

Double outlets, though, typically require something called a Rastelli procedure, which involves the placement of a conduit, or synthetic tube, that separates and redirects blood within the right ventricle. In this case the tube would be sewn over the VSD to direct all blood from the left ventricle directly to the aorta. The Rastelli's terrible drawback is that the synthetic tube can't grow, so this toddler will have to endure more open-heart surgeries in the future, first to replace it and then to replace its replacements. Also, these conduits tend to calcify over time, complicating both the heart's business and the surgeon's work, and have been known on occasion to shatter like eggshells—another risk.

But for the moment Shawna Cundiff and her husband, Edward, just want their baby to get better; they want him to stop turning that awful blue-gray color, and to grow. Shawna has had Patrick at home for several months and monitors his blood-oxygen level continually; he now has saturations of 60 percent. She got tired of looking at her blue son and took him back in to see his cardiologist, thinking, This just isn't right. "I could not give him a bath—you cannot put a blue baby in the tub," explains this skinny twenty-two-year-old mom with long brown hair and a pale complexion. With a three-year-old at home and Patrick perpetually ill, she hasn't slept much this year. Her cardiologist, who said a Rastelli on a seven-kilogram kid was too risky for his center, kept telling her to wait till Dr. Mee responded before they went ahead with the single-ventricular repair. "Everyone kept talking about Dr. Mee, Dr. Mee," Shawna says. "He's supposed to be world-famous or something. I never heard of him. I know Sylvia Brown, the psychic on *Montel*"—Montel Williams's talk show. "If he's so famous, why hasn't he been on *Montel*?" Skeptical though she may have been, she'd been glad to hear that Dr. Mee thought he might be able to attempt the Rastelli. After meeting with Mee, Shawna would say only, "He's kind of quiet."

The news among the doctors at the clinic the Monday before Patrick's surgery, however, was less optimistic. Each Monday morning at ten virtually the entire staff—surgeons, cardiologists, O.R. nurses, cardiac nurses, fellows, and residents—squeezes into the conference room, where first the upcoming week's surgeries and then the previous week's catheterizations are discussed, with echocardiograms and angiograms displayed on a movie screen at the front of the room (Dr. Mee's flat crown is generally silhouetted in the bottom left corner).

Echocardiologist Adel Younoszai opened the meeting: "The first case on the surgical list is Patrick Cundiff, born prematurely at thirty-three weeks with d-transposition." As he spoke, Geoffrey Lane popped in a video of the echo. The room, filled by about forty people, was quiet as the echo ran. Lane then screened, via computer, the angiogram from Patrick's catheterization. As these images rolled, Adel filled everyone in on the pertinent info—where the patient had been referred from, what happened with the previous shunts—and concluded with the question the referring cardiologist asked Mee: "Can he be a two-ventricle repair with a Rastelli?"

Mee, who'd been leaning back in his chair to watch the angiogram, a moving X ray of the blood flow through the heart, asked, "Are both shunts patent?".

Dan Murphy said he could see flow through one but not the other, and then added, "This looks like double-outlet right ventricle." (Such is the inexact nature of diagnosing, and naming, an infinitely variable heart defect—no one can quite decide what to call it.)

Mee asked, "How big is the VSD? Is it big enough?"

Cardiologist John Rhodes had already calculated the VSD at about eight millimeters. "For a Rastelli?" John asked, then answered, "No."

Lane let the angiogram run repeatedly through the various shots until Mee turned around in his seat to John Rhodes, at which point Mac, who'd been taking notes on individual cards, writing down what'd been discussed and the intended surgical plan, reached behind him to turn the room lights back on. Mee asked Rhodes, "Did you calculate the PA pressure? What if I can't do a Fontan?"

Rhodes reiterated that the pulmonary-artery pressure was a mean of thirty millimeters of mercury, but said he couldn't answer the second question. (A pressure of thirty is too high; in a Fontan, a single-ventricle repair, blood enters the lungs passively, and with excessive pressure, there won't be enough pulmonary flow.) What Roger was saying, now that they had all the data here before them, and their own echo and cath, was that he didn't think he could do a Rastelli *or* a Fontan: neither repair looked possible in this instance. Everyone was silent for a moment, and then Dan Murphy said, "Unfortunately, there are no great surgical outcomes."

Someone suggested doing a blade septostomy in the cath lab to generate more mixing as a fallback option, but Mee wanted to get into the child's chest and see this heart for himself. "The plan will be to cut away some of that tissue around the VSD and see what it looks like from there," he said, turning back to the screen and flipping to a new card. Adel said, "Our next patient is a three-year-old coarct . . ."—and the meeting moved on through two and half more hours and another dozen and a half cases.

After rounds the next morning, instead of returning to his office, Mee stops by the O.R. to check on how the opening of Patrick's chest is going. He tends to do this when a case is on his mind; this one bothers him because there doesn't seem to be any good solution. Emad Mossad is there, having anesthetized the patient, and he and Mee discuss the nomenclature of double outlet, which can range across a spectrum from, at one end, tetralogy of Fallot to, as in this case, transposition. "In a biological system, you have to use a nomenclature that embraces the full extent of the disease," Mee tells Mossad.

He leans back in the swivel chair at the O.R. vestibule's front desk, mulling. The doors open, and Appachi enters. Earlier, rounds got stalled at the bedside of one of his patients, who had stopped peeing; they couldn't figure out why and were worried about kidney failure, a not-uncommon result after the insult of bypass. Roger was hoping there was something wrong with the Foley, maybe a mechanical problem with the tube in the urethra, instead of a major organ injury. Now Appachi smiles and says, "Roger, we put in a new Foley, and she peed a ton."

Mee exhales loudly—*Ah!*—and says, "That was making me depressed."

"I thought you'd be glad to know."

Mee continues to grin, then advises Appachi, "Always look for sparrows before you look for canaries. Go back to the basics. Once you've exhausted those, then you can start hunting canaries."

Appachi returns to the PICU. He will tell me later that he knows Roger gets upset over such things, and he "didn't want him going into the operating room like that."

Roger sits for a moment, still thinking. Then he stands and walks to the O.R. door, punches the metal plate that opens it. Debbie Seim, the scrub nurse, is getting the setup table ready. "Hey, Deb?" he says. The name sounds like *Dib* in Mee's accent: "Hey, Dib, be prepared to do a switch in there."

Deb pauses, her eyebrows go way up, and then she repeats, "A switch?" Then tentatively, "All right." She's responsible for having everything he may need on the table, and no one's said anything about a switch—isn't this a double outlet?

Mee steps up to the table, as ever looking like an underwater explorer with all his gear—hood, mask, gown, gloves, loupes, headlight—and observes the heart he's been thinking about all morning. It's enlarged and bulky from the extra work and pressures it has had to endure. Roger immediately begins dissecting out the pulmonary arteries farther back, then putting the toddler on bypass and cooling him down to 22 degrees Celsius. This surgeon doesn't speak or bother with any kind of preamble when he gets to the table; his second foot is scarcely on the stool before he's in there with the tools, as if racing from the get-go. Patrick's heart is so overgrown with gristle from his two previous healings that Mee can't see any coronary arteries: the organ has the texture of something deep-fried. Mee clamps the aorta and directs Kevin Baird to run the cardioplegia into the aortic root to stop the baby's heart. The muscle pales and stills, gradually but quickly, finally collapsing. Mee slices a small hole in the atrium, then opens it wider with scissors. He pulls back gently on the opening of the tricuspid and peers down into the right ventricle. The echo showed excess tissue near the VSD, but no one could say how much, what kind, or whether it could be removed.

Mee says, "It looks like a goddamn *valve*."

The VSD is almost closed over, and he can't tell whether or not this tissue is part of the mitral valve. He has to enlarge a hole in the atrial septum so he can look through the left atrium and down into the left ventricle. When he's convinced that the valve tissue is extraneous, he removes it. Now he can enlarge the VSD and for the first time see more clearly the structures of this heart.

"The pulmonary valve is there," he says. It seems to be built into the left ventricle and appears to be functional. "The valve looks good," he announces, and then, referring to the VSD, "We can definitely patch this." He asks for a nine-millimeter probe and passes it through the VSD and then through the pulmonary valve. The valve is big enough, he decides; he's going to make it an aortic valve. "I think we can do a switch," he says, setting in motion not only a definitive biventricular repair but one that won't require a synthetic conduit.

He wastes not a moment in closing the VSD, using a Dacron patch and ten interrupted pledgeted 5-0 Prolene mattress sutures. These are long stitches with what look like tiny white cushions stuck in the middle—pledgeted—which prevent the stitches from pulling through the heart tissue. Mee runs each stitch through the circumference of the VSD and then through the Dacron patch, which he's trimmed with scissors to the shape he wants. When all ten stitches are in place, he's able to slide the patch down these guide-wires into place over the hole and tie off each of the ten stitches with rapid, gently secured slipknots. With a bulb syringe, a device that looks like a small turkey baster, he shoots water into the left ventricle to confirm that the patch doesn't leak. Next he inspects the tricuspid valve, which has been thickened over time by the jet of blood shooting through the VSD; the leaflets have become separated, so he puts in three separate stitches to reunite the valve tissue. Now he begins the actual switch, transecting the aorta, probing the coronaries with a tiny dilator; the gristle on the surface of the heart bulges as he snakes the dilator through to visualize the path of the coronaries. He finds that a right coronary comes off the left anterior descending and turns to cross the front of the aorta, while the right circumflex originates in the aorta and runs across the main pulmonary artery. But because each opening, or ostium, is centrally located, he doesn't foresee running into insurmountable difficulties in transplanting these arteries. "Let's be real careful here" is all he says before snipping them out of the aorta.

Mee transects the pulmonary artery, then performs a LeCompte maneuver to give the pulmonary artery more length when he attaches it to the aortic root. The switch proceeds perfectly, though the operation takes longer than he'd like. Also, he finds he can't get at the thickened tissue beneath the aortic valve without cutting more of the heart, and because the patient is in such good condition— heart beating well, excellent hemodynamics and blood gases—he doesn't want to attempt this additional repair, fearing that he might do more harm than good. He grows prickly only occasionally, when his assistants slow him down. *"Come on, Mac, picture what I'm doing,"* he says at one point; the words are almost inaudible, uttered in muffled italics. To Charlie, at his left, he says, *"Are you watching what you're doing?"*

Fackelmann asks Debbie, "Will you give Charlie a medium sucker?"

Mee says to Charlie, *"Why don't you ask for what you need?"* He shakes his head. "Debbie, I'm going to get Charlie some glasses he can see out of."

By the end of this operation Patrick will have spent three hours on bypass, and his heart will have been stopped for an hour and a half. With the right ventricle now pumping blood to the lungs and the left ventricle pumping blood to the body, with the coronaries filling well to feed the heart muscle, with pressure lines—wire-thin plastic tubes—inserted in the left atrium and pulmonary artery, with temporary pacing wires sewn into the heart muscle, with two rubber tubes punched through the skin below each side of the rib cage to drain the chest cavity, and with a Tenckhoff catheter inserted to drain the abdominal cavity or dialyze if necessary, it's done. Mee thanks his assistants and steps away from the table, having spent four concentrated hours on this case, leaving the closing to Fackelmann and Charlie. He signs the sheet attesting to the surgery and walks out of the O.R. After returning his headlight and loupes to their boxes in the cabinet, then removing the mask and hood and throwing both away, he mutters, "Not perfect." Then he heads back to his office, hoping Deb has gotten him some lunch. He's starving.

Mike Fackelmann works with Charlie on the closing because Charlie can be a little sloppy. Mike enjoys closing a chest. It can be a pleasure, especially when a case has gone as well as this one. He lifts the first foot-long 2-0 wire, on each end of which is a thick, semi-circular needle. Using a pair of powerful needle holders, an instrument that looks like a cross between pliers and snub-nosed scissors, he drives the needle up through the baby's breastbone till the tip pokes through the top, then grabs it again with the needle holders and pulls the wire through, leaving the holders still clamped on the needle. He then does the same with the other needle and the other half of the sternum. When he has five of these wires in, he and Charlie pull the chest closed tight and twist the wire by flipping the pliers from one side to the other. Mike then cuts the twist about half an inch above the bone and carefully bends the twist of wire back down into the sternum (if it's not completely flat, or a little twist is left sticking up, it may be a problem for the rest of the patient's life). He likes to close in three layers. First, using absorbable sutures, he sews a thin layer of muscle and subcutaneous tissue. Next he closes a second layer of subcutaneous tissue. (Because it's so easy to get a little buckle in one of the layers, especially the subcutaneous layer of muscle, and leave a crooked scar, he's meticulous here.) Finally he closes the skin, but in a way that no sutures are visible—a kind of zipper effect. Last of all he'll place some Steri-

Strips, small butterfly-style bandages, along the incision for additional support.

Mike hardly said a word all the time Mee was in the room, but his eyes were always darting around, as if he were *about* to say something. His gaze moved from chest to monitor, chest to monitor, chest to anesthesiologist, chest to monitor, chest to Deb, chest to the opening O.R. door, chest to monitor.

"I'm watching," he'll tell me, "because I want to know right away if there's a problem so Roger can stop doing that or we can adjust. I've got to watch because he can't, and if there's something not right, well, then, you don't have very long before you have to react to it—otherwise you've caused a little problem. One little problem leads to another. Little problems turn into bigger problems, so the more you keep those little problems down, the better the picture.

"I'm an assistant, I'm there to help *him* do *his* job," he goes on. "I'm not there to do perfusion, I'm not there to do anesthesia. I'm there to help him. That's my job, that's what I do. And if he couldn't trust me—and I told him this: 'If you can't trust me, then I have no business being here. If I can't do what you hired me to do, then what am I doing here?'—I can't do this fifty percent. I can either do it a hundred percent or I have no place here."

From early on Fackelmann fascinated me. Where, I wondered, had a working-class kid from the southwestern side of Cleveland come by this kind of intensity? Certainly not at home. I'd seen it before in obsessive chefs who were insane about details. Maybe Fackelmann *was* a kind of lunatic, actually crazy, but the kind of crazy that had fantastic benefits for patients. Later that year I would ask him if we could sit down and have a conversation. As always, he was a little surprised by the attention: What could I possibly want to talk to *him* about? He was just a tech. So I told him I wanted to know what made Mee such a good surgeon.

"Yeah, I can talk about that," he said. I knew him well by then and liked him; he knew that and felt much more at ease with me. Dr. Mee happened to be out of town that day, and because his open office was the only nearby place where we wouldn't be disturbed,

Mike and I sat down in there, at a small, round table in the middle of the general clutter. He would tell me what he thought about Roger, but as usually happens when a conversation goes on long enough, he would end up describing himself more than his ostensible subject. While everything Mike said was interesting, it was his relationship with his lawn at home that seemed to me to be the ultimate reflection of who he was.

"Roger's not fast, but he's got very good judgment," he began. On the Cundiff case Mee had pulled a switch out of what at another hospital would have been a single-ventricle repair; he did all kinds of switches, in fact, all year round. Mike cited this procedure as a good example of Mee's judgment: "Knowing how those coronaries are going to lie after you've sewn them onto the new aorta—when the heart fills up, you're gonna have the pulmonary artery pulling down on top of that. A good surgeon like Roger can think three-dimensionally like that, knowing how everything is gonna fit together, thinking about all those factors while he's doing each step, thinking three steps ahead about how one thing will affect another, concentrating not just on the thing he's doing but on the things he's *gonna* be doing. That's what makes him a great surgeon.

"Some people don't have that kind of judgment. You saw somebody here that had to be told, 'Look, hey, if you're gonna clamp the aorta like that, isn't this rotated?' I mean, you saw how it was, right? That's not something you can learn. You don't learn that type of judgment from anybody; it's something you're born with.

"You learn the mechanics of it, but every operation is different, there are variables in every case that you can't have experienced before. Say you've watched a hundred switches, and now you're taking that knowledge that you've learned from those hundred switches and you're going to do a switch. Well, the switch that you do will be totally different, so you have to use *judgment* in how to do that new switch. And the ability to use good judgment is not a learned thing."

I asked Mike about Mee's arrival, about what had happened between him and Cosgrove. He shook his head and said, "There's a lot of bad blood there."

I asked him why.

Mike said, "You know *what*? You don't get into a powerful role like this unless you have some ability to manipulate the situation. None of these guys are nice guys. You don't get to this level by being a nice guy."

I'd never been on the other side of Mee, but I had no doubt he could be an asshole in order to get his way, as he himself had said. Particularly in his dealings with the hospital bureaucracy, one of the clinic's trustees confided, he was considered "difficult." He'd always seemed pretty easygoing and decent to me, even in tense situations, and I said as much now.

Mike grinned and said, "I think he's the *best*."

"But you're not buddies."

"You let your guard down," he said. "You don't pay attention to what's going on. You don't respect people like you would otherwise—you don't fear them. Fear's not a bad thing. It makes people sharp, it keeps them on their toes. You want to make sure you do the right thing. Roger instills that in people. You see when you go on rounds how he manipulates discussions, almost playing the devil's advocate to get people to be a little more on their toes about what's happened. He'll say, 'I had him shittin' his pants.' He talks that way. It's a game, but it works, doesn't it? It's old-school, man. But he shouldn't be able to ask them something they can't answer if they're taking care of *his* patients. It makes 'em learn like crazy."

On the one hand fear was a good thing, Mike said, but on the other hand he didn't think of it as fear, or rather not a *nervous* kind of fear. "That's for amateurs," he told me. "You always do the same thing—you treat every case as if it's going to arrest." So in that sense anybody who was afraid probably shouldn't even be in there in the first place. "There's just one little tiny stitch holding that cannula in," he said. "How far is it in on a neonate? Maybe, what, three millimeters? You know how easy it would be to dislodge that? You do that, you get it back in, but now you've done that one little thing, had to compensate for a problem, and that leads to another problem. In a neonate that's a *big* thing.

"But you don't think that way when you're in there. You just have to make sure that every day you pay attention to what you're doing and do the best you can and pay attention to everything that's going on with the people in there and what they're doing. I *know* nothing bad's gonna happen. It won't happen; everything's gonna be fine. I'm gonna find out about a problem or somebody else will find out about it and we'll react the way we should react and things will be *fine*. You gotta maintain that attitude.

"If Mac and I get into the PA and I'm like, 'Oh, fuck, fuck, you just got into the PA, oh, fuck, what are we gonna do?!' then what's gonna happen? It's gonna be a disaster. I don't think there's anything that's gonna happen in there that we can't get out of. Like with Jonathan. He'd be unsure whether he could dissect something, and I'd say, 'I think it's good, just go ahead and cut that, I think you'll be all right. But if it's not, what's gonna happen? We're gonna have to put a stitch in it, and that's something you can do, right?'

"When Roger's dissecting a heart out of a redo, he's very quick, very precise, and he'll get into something. And he'll put a stitch in it. But you can't jerk around, you're gonna be there all day. So you gotta know that whatever you get yourself into, you're gonna be able to get yourself out of. Unless you cut a coronary on a . . ." And here he paused and issued a kind of chuckle, though not a pleasant one. "But that wasn't his fault. That time, that aorta, that thing was so small, that was—that kid did fine, that kid didn't die because of *that*." He paused and said, "But that stuff happens."

It was, he conceded, a serious business, but that was one of the things he liked about it. He believed he did a good job, and he wanted to be respected for that. He loved the precision of the work, the elegance. Clearly the precision was a part of who he was, and so I asked where he thought he'd gotten that from. He said he wasn't sure, knew only that his maternal grandfather had been the same way.

"I like to cut my grass and see the lines—perfect lines, like Jacobs Field," he said, explaining a summertime ritual he loved: cutting his lawn and then crosshatching it like a Major League ball field. He had a special mower, the kind they used on country-club greens and the like—a Scag forty-eight-inch walk-behind, cost him twenty-seven hundred bucks plus tax—that left excellent lines. "I'll actually stop before going the next direction to make sure that it's absolutely perfectly ninety degrees," he confessed. He laughed at himself, started to speak, laughed again, and said, "And people come by and they say, 'Man, your lawn looks really great.' But I don't need to hear it from them—if it's not right, I'll do it over again. My wife thinks it's crazy. And I do not want anybody walking on it until I can look at it for a while."

"*What?*" I asked. "No one can *walk* on it?"

"Not till I have a beer and look at it."

"Really."

"Yeah." He laughed and laughed some more, thinking about how much he enjoyed looking at it and also how odd it must sound. "Absolutely. It's obscene. Roger knows—I talk about my lawn all the time."

"Really."

"Some people aren't obsessive in their lives, but they're very anal in the O.R. Like Lytle"—another of the clinic's star heart surgeons, in adult—"he's the biggest slob you ever met, but I'll tell you, that guy is brilliant, as a surgeon he's brilliant. He's like Roger: he's always thinking, he's thinking three steps ahead of what's going on.

"Look at this," he said, spreading his hand out at Mee's office. "And *this* is *clean!* Man, I would be going crazy, pickin' stuff up." He paused for just a moment and then went on, "Ya know *what?* If you're not gonna do something right, then don't do it at all. That's my feeling. If you're gonna sew this kid up and you can't do a good job, I'm gonna take it down and redo it. You're not compromising anything by redoing it. But you are if the subcu isn't brought together evenly—you're gonna have a bump, the scar will be uneven. There's only one way to do things and that's the right way.

"Everybody knows how things should look, or how things should be done. But some people accept something less. Not everybody has that standard. Maybe the rest aren't skilled enough, or maybe they don't care; I don't know. A lot of people are like that about things in their life—this doesn't matter or that doesn't matter. Are those the type of people you want working on you? Or do you want somebody like Roger?

"I've seen that guy"—Mike exhaled brusquely, getting worked up now—"I've seen that guy, you're coming off pump and you're telling me that the mitral regurge is two plus, and you've got a *cardiologist* saying, 'I think this will get better.' But Roger's got enough common sense or good judgment to say, 'Ya know *what?* We're gonna recannulate and we're gonna go back on, and we're gonna do this over again, because when this kid leaves here, I'm gonna feel comfortable knowing that I did the right thing.' Instead of 'Hey, goddamn, it's ten o'clock, I haven't eaten yet, I really wanted to see this movie tonight,' or 'My wife's pissed off I didn't make it home last night. It's two plus? Ya know, I guess it'll be all right.'

"It's better to go back in and fix it, regardless of what time it is or how hard you've been working, because then you're gonna leave here with a clear conscience. Rather than have to come back the next day—which is what another surgeon who was here did repeatedly, you know, leave a residual VSD or residual ASD, argue with the cardiologist doing the echoes, 'No, that couldn't be there'— bring the kid back the next day and redo the whole operation, which is much more involved at that point.

"What's driving Roger? He wants it done *right*. Because he knows it's gonna bug the shit out of him if he leaves something that isn't right. It's gonna bug *me* if it's not right.

"Everything's a big deal, everything's got to be perfect. You cannot work in this . . . you can't *do* that. You just can't."

"Why not?" I asked.

"*Because*. This is a kid, this is *your* kid. Every little thing makes a difference. When these kids leave here, what do their parents look at, what do they see, every time they bathe that kid, when that kid grows up, when that kid starts dating? All they see is what they see here." Mike pressed a finger into his chest. "If you can't do a good job doing that, then you shouldn't be doing it. Because as far as the parent's concerned that's all that happened—that scar.

"I'll say, 'You know *what?* This is not good. I'm just gonna redo it. I'm sorry if you have a—no, no, that's all right.' A few years ago a fellow got real upset because I wanted to redo his closing. Usually I close anyway, but that day the fellow was closing, and I said, 'This is not good.' He said, 'Oh, it'll be fine, it'll heal up.' I said, 'No. You know *what?* That's not good enough, I'm gonna redo it.' He said, 'No, I'm telling you, it's all right.' 'No, it's not.' And I took it out and redid it. He got all ticked off, ripped his gown off and walked out.

"That's too bad, I'm sorry. If you can't do the right thing—I mean, this is a *kid,* this is not a Yugo here, this is somebody's *kid.* You don't take anything for granted, and nothing is *good enough.* It's got to be perfect. There's no compromise."

The following summer Fackelmann would receive what might be considered the ultimate praise. One day he was outside in front of his house when a neighbor from across the street approached him. The neighbor, after some small talk, got to the point: Was there any way he could get Mike to crosshatch *his* lawn, too?

Mike was honored. Absolutely, he'd love to do it, he said. The neighbor gave him a six-pack as thanks when he was through, but that wasn't the point, that wasn't why he'd done it. The neighbor asked if he could pay him. Mike said, "It's not about money, dude." Fackelmann loved the nature of the work.

8. "The Physical Genius"

The rounds scrum crowded into PICU room 11 as the resident, reading off a green sheet, said, "This is Patrick Cundiff, eighteen months old, status post–arterial switch operation, postop day one. . . ." The team listened to the presentation—a lengthy review, as this was the first one for the whole staff—in which the most salient information was that Patrick had been extubated the night before at midnight (indicating a speedy recovery) and that his chest tubes were still draining six to eight cubic centimeters an hour (the only troubling sign in an otherwise positive recovery pattern). The resident concluded with a rundown of the drips, listing the drugs the baby was on. Mee asked how much dopamine—a drug that helps the heart muscle contract—he was getting and then instructed, "Run that awhile—it was a ninety-minute clamp time."

Marc Harrison, leaning against the doorframe, said to me, "It's pretty amazing," referring to the arterial switch. "Roger really changed this kid's life."

I saw it happen on three other occasions that winter: a sick infant arrived at the clinic—in one case the referring doctors were so hopeless they'd been able to offer only a transplant—and Mee pulled off a biventricular repair. These cases described Mee's excellence as a surgeon better than any statistics could. And in any event, postop care had gotten so good that mortality figures could vary considerably in their meaning. Surgical mortality was defined

as any death within twenty-eight days of surgery. What that defini-
tion didn't convey was how sick a patient who *hadn't* died might've
been during those twenty-eight days and beyond: was she still on a
ventilator thirty days postop, did he die a year or two later, did he
require a second or third surgery, would she be able to add and sub-
tract after the long circ arrest, how healthy was he once he left the
hospital? Thus, the only figure that really meant anything was a
bad mortality. There was no excuse for that, it seemed to me; centers that had a crummy overall mortality, say above 10 percent,
probably shouldn't be offering surgical repair at all, given that most
places do so much better—and here, "better" meant more live kids.
But if the very best centers were under 2 percent, what did that
mean for the ones that were just *under* 10? Should they, too, pack it
in? Then again, a center could have a great mortality but a lot of really sick patients—no one necessarily monitored that.

The average length of stay in the ICU was perhaps a more
meaningful statistic—did kids recover quickly and get sent to the
ward, or did they linger in the ICU? The length of time to extubation was yet another indicator of how fast kids healed. A third
index of success in a pediatric heart surgery center was its percentage of reoperations. Patrick Cundiff had had a shunt put in as a
neonate to deliver more blood to his lungs. Mee, too, tended to opt
for the shunt when he could, thus creating, in effect, a false ductus
made out of a little white plastic tube, inserted in a closed-heart operation performed not through the chest but instead through the
side, between the ribs (a type of opening called a thoracotomy).
This way the child would have adequate blood mixing and could
have definitive surgery when he or she got bigger, when the myocardium was mature, had turned from jelly to firm muscle, making
the repair easier and better, and when the body was, in Mee's opinion, less vulnerable to the havoc wreaked by the bypass machine.
(There remained fierce debate over the effectiveness of the shunt-
and-wait philosophy versus the early-and-complete-repair approach so vigorously advocated and practiced by surgeons at
Children's in Boston and elsewhere, who believed that heart and
lung development was improved when everything was fixed as
soon as possible. This strategy furthermore eliminated problems re-

lated to the shunt and the risks of a second surgery, not to mention reducing the financial and spiritual cost to the family. Finally, the early-and-complete guys argued, a neonate tolerated bypass better than an infant.) But Patrick had had to be shunted a *second* time at his first hospital in an open-chest procedure. Had he really needed two shunts and then a repair—that is, two open-chest operations, the second of which was especially risky and difficult? Maybe or maybe not; it depended on where he happened to have his surgery.

In Cleveland Mee, with the help of preoperative and postoperative teams, had been able to perform on Patrick a small miracle of ingenuity and craftsmanship (ingenuity based on experience and innate smarts, craftsmanship developed over thirty years of almost daily practice). But it had to be more than that, I kept thinking; others were smart and had worked just as long. What *was* it that made Mee so good, why was he consistently ranked by his colleagues among the top five congenital surgeons in the world? And why had Dan Murphy said, "Roger's repairs look like God did them"?

Pediatric cardiac surgery was one of the most technically difficult specialties in medicine; in no other specialty did a patient's life hang so plainly in a life-or-death balance, ready at any instant to be tipped one way or the other by a single surgeon's hands. What made Roger Mee so good at this unusual work?

In May 2001, a couple of months after the Cundiff case, I attended the World Congress of Pediatric Cardiology and Cardiac Surgery, a once-every-four-years gathering of thousands of docs, nurses, and technicians (and sales reps), held that year in Toronto. On the third morning of a conference during which multiple seminars, symposiums, and plenaries were scheduled simultaneously, one event was standing-room-only, in an enormous room filled to capacity with perhaps a thousand people. It featured Mee and Guillermo Kreutzer of Argentina (an influential surgeon in international cardiovascular surgery, credited with originating, along with François Fontan, the Fontan procedure), Ed Bove and Marc de Leval, and the elegant, white-haired Aldo Castañeda—together "not only the greatest surgeons in the world today," said Welton Gersony, the lone cardiologist asked to join the group, "but ever." Before the session, as I waited in the pack to say hello to Castañeda, whom I'd met in Cleveland, de Leval said, "It's like trying to see the pope." This was a gathering of the superstars, the surgical legends, the men who walked on water, and they commanded a packed house. If there were a way to measure pounds of ego per square inch, I'd reckon there wasn't a concentration like this anywhere else on Earth.

De Leval spoke on the topic of one of his great concerns: the element of human error in what he called the "careful violence" of the O.R., which he likened to a cockpit, an enclosed space for fighting. Human errors were inevitable, he suggested, and so the O.R. must have an "error-tolerant system." He hoped one day to make an error-management expert—someone formally trained to spot errors and fix them—a fixture in the O.R. De Leval identified two types of errors, minor and major, and his most trenchant finding was that in procedures resulting in death or near-death due to human error, catastrophes were more closely linked to multiple minor errors—the kind with little noticeable negative effect—than to a single major one. It was those minor errors that went unrecognized and uncompensated for, in other words, that were most likely to bring about, three steps later or ten steps later, disaster.

Indeed, Mee had explicitly told me that one of the things he felt he was good at was spotting potential disaster long before anyone else could see it coming—recognizing those minor glitches and anticipating how they would inevitably cascade into major ones if they weren't corrected at once. He seemed to have the whole operation mapped out in his mind, and if it began to diverge from the imagined version in his head—a picture of what he'd envisioned happening held over what was *actually* happening—he could compensate or jig it back onto its track before proceeding.

Next Dr. Castañeda, introduced as the man who had proved wrong the prevailing dogma that neonatal surgery necessarily carried a high mortality, spoke about his pediatric cardiac center in Guatemala, established after his retirement from Children's Hospital in Boston. Picking up on de Leval's remarks, he noted that if he and his colleagues there had compensated for every potential error in the operating room, they would never have gotten a center started in that third-world environment. That was how far behind they were in Central America, he said, relative to the industrialized countries. It was a struggle; he had done approximately six hundred operations, with a 14 percent mortality. His immediate goals were to strive for a lower mortality, to properly train three additional pediatric cardiac surgeons to carry the center forward after his definitive retirement, and to offer training opportunities to surgeons from the other Central American republics and the Caribbean.

The last speaker to take the podium was Ed Bove, a tall, long-limbed man with a narrow face who had been asked simply to reflect on the facts and fantasies surrounding the peds heart surgeon. Born in New York City, Bove had trained in Albany and done his transformative learning under de Leval at Great Ormond Street in London.

I loved listening to these guys. The responsibilities and consequences of their work continued to amaze me, and Bove in particular struck me, both in his presentation and in a few subsequent meetings we had there and at his center in Michigan, as a humane and gentle personality. He was also unusually articulate about what he did. He did see it as an art—"Art and science seasoned with skill and decisiveness," he'd told me the day before, quoting from his talk. "You have to be able to make decisions quickly, and they have to be the right decisions most of the time."

For the standing-room-only crowd he reviewed the history of the specialty, noting how young it still was, and how critically surgeons today were studying the results of their operations. Surgeons didn't cure patients; rather, they took acute life-threatening conditions and turned them into chronic conditions that were easier to manage. Surgeons were usually the last hope in a patient's quest for improvement, and they typically shouldered that responsibility alone—as a consequence of which, he said, "some of us might suffer delusions of grandeur and be unable to see ourselves as others see us." But he also noted that the qualities associated with the heart surgeon—"persistence, tough-mindedness, strong ego—are precisely those traits required to operate inside the human heart with a life in the balance." Later he would add that "ego and personal challenge have no role in surgery."

He wondered aloud to the crowd, Is the lowest-risk procedure always done? "Sometimes ego and an eye toward the big picture get in the way," he answered. "We can all remember times when we perhaps bit off more than we could chew under the guise of achieving a magical complete repair, when in reality it was an unwillingness to accept real or perceived limitations either by ourselves or our colleagues.

"Congenital heart surgery," he concluded, his voice echoing in the darkened hall, "is special. It's a young discipline and a unique specialty. It's a profession that combines art and science seasoned with skill and decisiveness. It's been characterized by industry and innovation, and in part because of its limited numbers, it still attracts some of the best and brightest in medicine. Cardiac surgeons are persistent and tough-minded, though a shortage of ego has never been a trademark among heart surgeons, but it may be argued that it's that very ego that has played a pivotal role in the growth and development of our specialty."

While these remarks painted an accurate portrait of the work, they only brushed the surface of what made Bove himself, or Mee, so good. This was harder to get at. One writer who tackled this elusive question was Malcolm Gladwell, writing in *The New Yorker*.* A reporter of ideas, Gladwell used the term "physical genius" to try to limn the qualities that made a performer stellar—what separated Michael Jordan, or Wayne Gretzky, or Yo-Yo Ma, or the vaunted neurosurgeon profiled in his piece, from their excellent colleagues? What made them so clearly better?

Gladwell conceded that it was tempting to regard the physical genius like an intellectual genius, whose excellence could be defined by a single factor, an IQ. He identified several factors, such as "force regulation," meaning how hard to press or pull—an intuitive sense that in the physical genius might well be highly refined and, in terms of neurophysiology, lightning-quick—which was critical

---

*"The Physical Genius," *New Yorker*, August 2, 1999.

for the pianist, the neurosurgeon, or the peds heart surgeon who must join a four-pound neonate's aorta and pulmonary arteries, delicate as wet tissue paper, while racing the clock.

Physical geniuses might also have a more highly developed pattern-recognition mechanism, Gladwell wrote: "What sets physical geniuses apart from other people . . . is not merely being able to do something but knowing what to do—their capacity to pick up on subtle patterns that others generally miss." This might be another part of that "feel," the unidentifiable set of qualities that even the surgeons and athletes and musicians couldn't articulate, the ability to sense just what to do and when to do it.

Gladwell identified practice—not the love of it, necessarily, but simply the capacity for it, the ability to do the same thing over and over again—as a commonplace among those who might be said to have a physical genius. These people were obsessive.

Another feature of such talent was a gift for what psychologists call "chunking," the gathering of a series of single facts into one fact. It was the way the rest of us remembered phone numbers, the way a chess master could look at a game in progress and then re-create it on an empty board, because the dynamics of all the pieces in the game at that moment were, to that master, a single fact.

"When psychologists study people who are expert at motor tasks," Gladwell continued, "they find that almost all of them use their imaginations in a very particular and sophisticated way." He cited a surgeon who, "when he gets into uncharted territory in operations . . . feels himself transferring his mental image of what ought to happen onto the surgical field." Mee seemed to use the same skill, as he had explained it to me, to avoid the small errors that would cascade into major problems if they weren't compensated for.

Part of imagination was visualization, and here Gladwell referred to the work of Stephen Kosslyn, who has shown that this act consists of several pieces, including the ability to generate an image in one's mind, to hold it steady and to rotate it, and to be able to inspect its various aspects.

"If you think of physical genius as a pyramid, with, at the bottom, the raw components of coordination, and, above that, the practice that perfects those particular movements, then this faculty of imagination is the top layer," Gladwell wrote. "This is what separates the physical genius from those who are merely very good."

All of the above applied to what I'd observed in Mee. In this specialty, with all its reconstructive requirements—putting in shunts and patches, fashioning entirely new vasculatures (as opposed to neurosurgery's typical excisions—say, of a tumor coiled within the spinal cord)—the peds heart surgeon had to rely on advanced imaging, the ability to generate an image in the mind and manipulate it. Whenever Mee put a patient on bypass, the heart, a big, pulsing bulb, would deflate, go flat, as would the major vessels coming out of it. He had to perform all his work by inflating that heart in his mind in order to create new structures in or on it, structures that must also be flat while he was cutting and stitching—that is, he must connect them flat, but they must have the capacity then to fill up with blood, and the blood must flow freely, without obstruction.

Gladwell guessed that it was not only artists such as Yo-Yo Ma who derived some sort of aesthetic pleasure from their imaginative improvisations, but also people like his neurosurgeon and perhaps Roger Mee, who loved, as he'd told me, "an ASD that goes like poetry." Or perhaps it was that capacity simply to think through in his head what a case would look like, then stand, scratch his belly, open the O.R. door, and call to his scrub nurse, "Be prepared to do a switch in there."

Chuck Fraser, one of Mee's fellows in Australia and then later his partner at the clinic, while not wanting to pull the water out from under his feet—"His talents are immense," Fraser said—believed that technically Mee might not even be the best. "There are a lot of really smart cardiac surgeons, and there are surgeons who are technically more artful—but their results aren't as good." What accounted for his superlative results, Fraser said, was that "he is doggedly persistent. He is absolutely determined to get it right on every patient."

Though he arrived in Melbourne with a Johns Hopkins attitude, he said, Fraser recognized from the outset that if he wanted to do congenital heart surgery, Mee's system, Mee's methods, and Mee's attitude and drive were the only way to go. The ultimate proof of Mee's excellence, he said, was that he, Fraser, was now leading a center whose results were every bit as good as Mee's across the board (he and his partner performed about seven hundred surgeries a year, with an overall mortality consistently below 2 percent, he says).

Mee was no better at describing his own particular genius (a term he'd feel uncomfortable applying to himself), and probably worse. This kind of discussion, when you got down to it, he considered to be "wanking." When he talked about surgical excellence, a subject that intrigued him, he was more apt to speak of details, the little things. "It's amazing how few people know how to do the basic things," he said. "Like having a clear field or knowing how much tension to put on a stitch." These things were vivid to him, as they were to all excellent surgeons. When a surgeon *had* it—the gift, that natural excellence, that touch, the grace—Mee said, "it sticks out like dogs' balls."

He continued, "I try to mimic nature. It's a superb design"— the heart and the body's vasculature—"and I've got tremendous faith in the engineering that nature's come up with, I think it's brilliant. . . . When you're doing a shunt, it's important not to introduce a new abnormality. I think that in general [the success of the surgery] has a lot to do with how much distortion we create. Is it the right length, is it in the right position? If you want to take a Gore-Tex shunt to the right pulmonary artery, it should be placed exactly on the top edge of the pulmonary artery, not on the front, not on the back, because that's going to twist it. It should not be pulling the pulmonary artery up or pushing it down, so the length has to be right. If you look at angiograms anywhere in the body, it's a neat system, the branches take off at the correct angles, they're not kinked, so I try and mimic nature as much as possible. I don't think we can do better. We can do better than nature gone wrong—that's our job. But nature gone right is something to try and emulate."

Ultimately he diminished what he did in the whole scheme of

things. "It's very easy for a person to focus only on what happens in the operating room. That has to be done well, expeditiously, and you have to do the right thing, the right procedure for that patient. But if you focus only on how neatly Dr. Mee sews, well, I take that for granted. There's no excuse not to. If you can't do the hand stuff, you shouldn't be in surgery.

"But because the craft is just a little bit of it, it's easy to say, 'Surgery's just a little bit of it, so it doesn't really matter if I'm a good sewer or not.' Everything matters. *Everything* matters. The weak link can be anywhere.

"Transporting the patient to the ICU, the operation can go well and then hell can break loose"—de Leval, in his discussion on human error, had noted that such errors often happened on this trip—"so it's easier to do if you have a short distance to go. It's no-brainer stuff. It's self-evident. Each little thing is a no-brainer. It's hard to convey to the people you have to negotiate with to put this all together, the importance of every link in the chain. Each person has their little bits and pieces, all those tiny decisions. But they get fouled by personalities, ignorance, lack of diligence, sometimes overdiligence—by everything that's human."

9. The Good, the Bad, and the Good Enough

As the scrum moved out of Patrick Cundiff's room, I asked Dan Murphy for some perspective on the case. The questions raised by Drew H.'s case remained vivid in my mind. Had the referring cardiologist been justified in offering Patrick's parents only a single-ventricle repair at his original hospital when, as we'd just seen, a biventricular repair was possible?

As rounds proceeded to the next patient, Murphy said Roger had done a "brilliant job" and assured me that the referring hospital's diagnosis and surgical recommendations had both been reasonable. How, he asked, could someone recommend something that wasn't in his repertoire? Dan reminded me that even in all the discussion here, there had never been any mention of an arterial switch. "I wonder how many cases like this *Roger* has even seen," he said. "This was just creative thinking."

Having watched the operation and stayed for the closing the previous day, I'd gone to Patrick's ICU room shortly after his return. It's always a time of busy activity: the simple act of rolling a sedated child the mere hundred feet from the O.R. to his or her ICU room can't ever be taken for granted, as more than one person explained to me, because so many things can go wrong in a patient who's cold, who's being ventilated by hand, whose heart has been manhandled, whose blood has been flowing for a long period through many feet of plastic tubing, whose life depends on the continuous delivery of exact amounts of powerful inotropes, and who may be bleeding inside. Immediately the room is filled with purposeful doctors and nurses hooking up monitors and putting the patient on a ventilator, checking all the while to make sure he or she isn't on the verge of crashing. Appachi was the attending that particular day, and he oversaw the others as they went about their work. When the routine was nearly complete, only Appachi, a respiratory therapist, and a nurse remained in the room. Patrick's mom and dad, Shawna and Edward, entered, expressionless, and made a beeline for his bedside. Appachi stepped to the opposite side of the bed and said, "I'm Dr. Appachi. He's doing beautifully." He didn't need to say anything else, other than that he'd be here to answer any questions they might have. Then he left the room to let them spend a few minutes alone with their child.

Outside in the corridor Appachi said to me, "It happens all the time: parents are told that their kids can't be fixed, and they come here, and Roger fixes them."

And now I asked the question that had been running like an electrical current beneath everything I'd witnessed: Why hadn't Patrick been sent here in the first place, given that the cardiologist obviously thought enough of Roger to refer the kid to him later on, when he ran out of options? For that matter, why didn't Mee do every switch in this part of the country? There were, I knew, three other hospitals in Ohio alone with congenital heart centers that did the arterial switch; one of them was even within walking distance of the Cleveland Clinic. But how could any cardiologist in the state send a baby with transposition to anyone other than Roger Mee if he or she truly wanted to do the very best for that patient?

I was expecting a balanced explanation from Dr. Appachi, the kind of careful approach most physicians took when addressing a tricky issue, a response that noted the many factors that might be involved, the number of annual cases, the difficulty of traveling, insurance tangles, issues of hospital affiliation and ingrained referral patterns.

But Appachi said, "I don't know. I think it's child abuse."

· · ·

Patrick got out of the PICU a couple of days later (the average stay for an infant who'd had an arterial switch was thirty-six hours, according to Appachi), and I took the opportunity to speak with his mom, Shawna, whose husband had already returned to their hometown and gone back to work.

An intelligent, loquacious young woman, Shawna described her baby son's tortuous course: the air ambulance from the community hospital after his premature birth in July 1999, the three weeks in ICU, the emergency shunt procedure after the duct closed, then more weeks in the ICU. It got scary, she said, when they sent the chaplain around to see her after the surgery: "I thought, My God, is it that bad?" Shawna recalled. She and Edward were finally able to take him home, but a few months later he started turning blue and required a second shunt via an open-chest operation. Again, he was in the ICU, this time for about a week, and again her days were punctuated by creepy visits from the chaplain. "It was awful," she said. But again Patrick healed and came home. About a year later, though, his color changed, so she took him to the cardiologist who set in motion this latest round of treatment.

Shawna was surprised that Patrick had gotten out of the ICU so quickly—he'd previously been such a slow healer. As Mee had noted, though, there was no such thing as a good healer or a bad healer. Now she wasn't sure *what* to think—"I don't know if they don't know what they're doing there or if they don't know what they're doing up here!" she said.

A week later I called her at home to find out how Patrick and the rest of the family were doing. Patrick was napping and doing well, she said, and sleeping through the night, so she could at last get some sleep. She'd already taken him to see the cardiologist. The whole experience seemed only recently to have sunk in for her: "I was really shocked," she could say now that she was home. "I told his cardiologist, 'I don't know what happened, but he did so good up there. Usually he does real bad, like when he was down here and had the surgery.' The cardiologist said, 'You have to understand, he was with one of the top surgeons.' He was surprised Dr. Mee did the correction surgery. He couldn't believe how pink Patrick was. All his color is excellent."

Shawna said she'd rather have visited Cleveland to go sight-seeing, but all in all she was elated. "You know it's not the end," she said before getting off the phone, "but you can rest easy for a while." Mee had predicted that Patrick would need surgery again at some point because some pulmonary stenosis remained, but a year later the boy would be doing so well that no surgery seemed to be required, at least in the foreseeable future. "He's doing great—in day care and running around," Shawna would tell me, adding that to look at him, no one would know there'd ever been anything wrong.

I knew Mee couldn't possibly do every switch in the United States: no firm numbers are kept, but transposition accounts for about 10 percent of all congenital heart defects, making it the second most common type of defect. Extrapolating from known data, that's about 1,250 to 1,600 babies born with TGA each year. I knew that the next generation had to be trained, that the system had to replenish itself. And I knew that not everyone could be one of the world's best. But what bothered me most was the notion that as a layperson, I likely wouldn't have had any choice in the matter if my child were sick: because hospitals refer patients to their own except in dire emergencies, had one of my children been born with transposition, I'd probably never even have heard Roger Mee's name. The hospital where my kids were born, a facility I could see from the clinic's windows, had its own surgeons, so I'd have been referred to one of the two there, neither of whom had Mee's experience or results. If I'd demanded that my cardiologist—a doctor I'd have gotten by chance, simply because he or she was on call that day, another of those luck-of-the-draw situations inherent in the medical system—get me a second opinion or offer me more options or find me a more experienced surgeon, I may well have been sent to Michigan, or Boston, or Philadelphia, all big centers with good results, but requiring substantial travel and time away from home. There was—and had been since the founding of the one, in the 1920s, by renegade doctors from the other—a universally acknowledged rivalry between the Cleveland Clinic and University Hospitals. Before a UH cardiologist would refer a patient to the clinic, just a few blocks away, he or she may well have referred to the University of Michigan's heart center run by Ed Bove, a surgeon whose stature and results were comparable to Mee's but whose O.R. was a three-hour drive away. (When I visited Michigan, a circulating nurse lowered her head and looked up at me, a gesture of surprise that was not really surprise, and said, "We get a *lot* of referrals from Cleveland.")

"A lot" of other, unnamed patients was one thing, but *my* kids were another matter altogether, and like any parent, I could think of every patient only as if he or she were my child. The year I was a privileged visitor at the clinic, the competing center at the hospital where my kids had been born performed four switches, with one death, according to the state's Bureau for Children with Medical Handicaps, a division of the state Department of Health, which gathers such data. The clinic, by contrast, had done seventeen, with no deaths. But I wouldn't have known the difference or been offered the choice; and had the problem been one that required prostaglandins, as was the case for Connor Kasnik, or one requiring immediate surgery, I'd hardly have had time to research the subject, given its complexity and my own inevitable emotional disorientation.

"The parents don't know," said a clinic cardiologist named Tamar Preminger, who added that the differences were particularly dramatic in this specialty because the surgery itself was so difficult and the stakes were so high. "Thousands of people can take an appendix out," she said, "but there are not thousands of people who can do congenital heart surgery."

I couldn't let it go, though it was almost too scary to think about.

Anesthesiologist Paula Bokesch, dressed in white pants and shirt and a white lab coat, caught Roger Mee at the scrub sink. He was running the yellow-bristled brush over arms and elbows splotched pink by psoriasis, his hood, mask, loupes, and headlight all in place. Paula said, "Roger, I'm ready to bring her in. Have you talked with the family yet?" The patient, S., a one-and-a-half-year-old girl, had arrived at one o'clock this morning by air transport.

"No," Mee said. "I meant to between cases, but I've sort of run out of time."

"I can't wait," Paula said.

"I went round, but they'd gone off." Mee shook his head in frustration. They needed to begin anesthesia, but no one had told the family that their child's operation would be today; it was a complicated case, the girl very sick from a botched job at another hospital and now with numerous complications, and a slot had opened up at the last minute. Roger made the decision at the scrub sink: "The family brought her up here for *surgery*. Bring her in and tell them I'll talk to them before I go in. We're going to do a left thoracotomy and fix the arch, and if the hemodynamics are stable, we'll do the repair."

Paula went to look for the family. Roger fell back against the metal plate on the wall to open the door, then called to Fackelmann, "Do a left thoracotomy on the next case."

Fackelmann, then assisting in opening a twenty-six-year-old man on the table in front of him for an elective VSD closure, said, "You think you can get to the arch from the left?"

Roger said, "Yeah." Mike shook his head, but he didn't say anything.

Mee had first met the family hours earlier, when the scrum rounded in the girl's room. S. had been born in July 1999, the same month as Patrick Cundiff, with tetralogy of Fallot (a VSD right below the aorta, pulmonary stenosis, and, as a result of these, a stiff, enlarged right ventricle). Among the most common of heart defects, tet is routinely repaired and, in straightforward cases, should have almost no mortality. S. *had* been repaired, but instead of getting better, she'd stayed in the ICU and kept getting worse until her cardiologist, who had been a fellow at the clinic before moving west, referred her to Cleveland for more surgery. An initial echocardiogram was performed shortly after she got to the clinic, and the problems noted by the referring hospital were quickly confirmed:

the VSD patch the first surgeon had put in had come undone, and there was obstruction in the left pulmonary artery. Also confirmed was another defect that the previous hospital had identified weeks *after* her surgery: S. had a double arch, meaning that her aorta divided, split around the trachea, and came together again on the other side. Left unrepaired, it was now squeezing her airway.

When the group had finished rounding in S.'s room, Mee stepped to the side to speak with Nina, the baby girl's young mom, and her parents. They were blank-faced but focused hard on Mee. He said, "There are two issues here, the tetralogy and the airway. The double arch has formed a ring around the airway, and I'm trying to decide if she needs two operations. Do you understand?" The granddad nodded, but slowly. "The good thing," Mee continued, "is there are clear reasons why your kid is sick. If there were just a lot of little details, and she was this sick, then . . . But we know *why* she's sick. There's a hole in the heart, which we can fix, which is part of tetralogy. And we can fix the double arch. But it's formed a ring around the bronchus and there's a lot of cartilage there, so it's bound to be a little mushy—so even if we fix the problem, I can't guarantee we can get her off the ventilator." He waited for any of them to ask a question or indicate some comprehension, but all of their exhausted-looking faces remained blank as stone as they just continued to nod their heads, slowly. He couldn't know how much of this they'd taken in. He said, "I'll come back and talk to you after rounds."

S. lay on her back. Growth hadn't been a problem: she was unusually large for her age at thirteen kilos, nearly twice the weight of Patrick Cundiff. And she seemed even bigger than that because of severely edematous, or fluid-filled, tissue, a result of the previous surgery and bypass, of being on the ventilator, of her airway obstruction and bad heart; her whole body was swollen, as if she'd been overinflated. She was sicker than anyone had anticipated and needed surgical repair *soon,* but Mee had two or three surgeries scheduled every day for the rest of the week. He'd have to operate today. When he saw Dan Murphy and Geoff Lane in the carpeted office corridor after rounds, he asked them, "Does anybody see any reason *not* to operate today?" No one said no.

Mee asked Lane to pull up the angiogram so they could have another look at S.'s strange arch. They found an unused computer in the echo lab, and with a few clicks of the mouse they were watching the moving X rays of blood flowing through the double arch. Mee wondered aloud, "Where does the right subclavian come off?"

Dan Murphy said, "If it's a right arch, it should come off the back of it."

"But it's a left arch, which is very unusual," Mee said. "I've never seen it."

"How easy would this be to miss?" Dan asked Roger. The previous surgeon, after all, had been *inside* this chest; one would think he'd have noticed a double aortic arch.

But Mee said, "It would be easy to miss. They would say, left arch with a right innominate." He continued to study the angiogram. "But we're all convinced that the right is the diminutive side."

Dan looked down at the beeper on his belt. "Two different operating rooms are calling me," he told Mee, "meaning pretty soon one of them will want you."

Mee returned to the PICU to look for the family again, but they'd gone downstairs to get something to eat. He asked Marc Harrison and Steve Davis, the two attendings, the same question he'd put to the others: "Does anyone see any reason not to operate today?" They both shook their heads, and it was decided. Mee spotted Kim Teknipp, the preop cardiac nurse, and told her he would operate on S. next case to try to fix the arch.

"If we're having a lot of trouble and it's a bumfuck," he said, "then we'll stabilize the kid and come back."

Kim nodded—she'd find the family and tell them what the plan was. Roger strode out of the PICU, headed for the O.R., where he would close the twenty-six-year-old's VSD.

Two hours later, in O.R. 51, Mee stared into the center of S.'s body. She lay on her right side beneath the green draping. Fackelmann and Mac had performed a left thoracotomy at Mee's request (the previous surgeon had put the shunt in through the right side, and

Mee wanted to avoid having to dig through all that scar tissue), making an incision between the fourth and fifth ribs, opening the intercostal muscle, then retracting the ribs. Mee worked through this hole on vessels that pulsed several inches beneath the surface of the baby's skin. He was talking a lot more than usual: "This is weird. . . . This is bizarre. I don't like this at all. . . . I'm not sure *what's* going on. . . . What the hell is that? . . . Oh, Jesus. . . . We've got some real bizarre stuff. . . . Oh, boy. . . . Oh, boy."

Then, "Oh, shit. Get me a seven-oh double-handed stitch. Come on, come *on*, have this stuff on the table—you're bound to need it." He put the stitch in and soon asked, "Can we get a cardiologist in here? This is an emergency. Can you get Dan in here? I'm beginning to worry about anomalous venous return."

Paula, on anesthesiology, was having a difficult time of it, too; unable to ventilate properly, she'd begun to bag by hand.

Dan Murphy arrived promptly, and Roger tried to describe the unique vasculature he'd encountered. Mee asked Murphy to reexamine the echocardiogram to make sure the pulmonary veins, the veins returning blood from the lungs to the left atrium, were working correctly. Dan left again, and Roger meanwhile repaired a hole he'd accidentally made in the esophagus while trying to repair the arch and remove the ring of cartilage that had formed around the trachea and esophagus. When Dan reappeared, he said, "Roger, the veins are normal. She *does* have a left SVC." Mee continued working without looking at Murphy. He wouldn't be able to do the arch repair from this side after all.

"Roger, why did you do a left thoracotomy?" Paula asked.

"I didn't want to burrow through all that scar tissue. I was hoping to do it here, but obviously I made the wrong decision." To Makoto Ando, he said, "It looks like you're going to have to do another unpleasant opening on the right."

Mac grunted, nodding, and chuckled nervously. He'd been on call the night before. I guessed he must be pretty tired by now. Fackelmann said nothing. Mee would close the left side, leave the O.R., and come back once Mac and Fackelmann had flipped S. onto her other side and reopened. The arch repair would go smoothly after that, but Mee would be unable to do the full repair.

Mee made room in his schedule for the definitive surgery by the end of the week, and this time the procedure was completed without incident. He resewed the old patch over the hole in the ventricular septum with pleageted stitches. He removed the patch that the previous surgeon had used to enlarge the pulmonary artery and sewed in a conduit that would allow blood to be pumped smoothly out of the right ventricle, through the pulmonary arteries and to the lungs.

Whenever I observed an operation, I always stood on a stool at the head of the table next to the anesthesiologist, watching over the cage. I often had a clear view of what Mee was doing, but sometimes he worked on a collapsed heart too deep in the chest for me to see. I'd wanted to get a look at the earlier repair but had been unable to. After the surgery I followed Mee to his office. Debbie had pushed aside the papers on his desk to clear a place for a cardboard tray that held his cafeteria lunch. I asked him how it'd gone.

"I think it was quite a good repair," he said, tucking into a six-ounce burger piled high with grilled onions, bacon, and cheese.

I asked how the first repair had been.

He finished chewing a mouthful. Then he hesitated some more. "It wasn't excellent," he said, smiling in a way that made me uneasy; he was bristling either at my question—was I looking for dirt?—or at the mental picture of what the previous surgeon had done. It was probably a little of both, but he elaborated anyway. He said the VSD patch had been sewn using a continuous stitch. The problem with that technique was that if one part of the stitch came loose, as it had done here, in this turbulent part of the heart, "it's like bridgework coming out," he said, squeezing his teeth. This particular VSD was in one of the most inaccessible parts of the ventricle, making stitching more difficult than usual. Even so, the surgeon had also cut too big a patch for the pulmonary artery and then had sewn it around a bend in the artery. "Patches don't turn corners," Mee explained. This one had instead buckled inward, obstructing flow to the left lung rather than facilitating it.

"And she had an obstructed airway," I recalled.

"I forgot about that," Mee said. "No wonder the kid was so sick." *Chomp*—the burger was almost gone.

. . .

I didn't want dirt; I just found the situation so disturbing that I had
to ask for details. I'd already had a talk with Nina, S.'s mom. A gentle woman with a pretty, round face, brown hair, and a sweet southern accent, she seemed to me too young, at twenty-one, to be going
through something like this. Several days after the final surgery I
passed her as she was leaving the PICU and asked how she was.

"I just want to go home," she said, looking sad and tired.
"Some days are OK. Some days I'm depressed. But I've been doing
this for eight weeks now, so I'm used to it. Yesterday was a good
day. She got her tube out, and I see getting her tube out as being
one step closer to going home."

Our first conversation had made me more uncomfortable.

This was what she'd told me: two weeks before her daughter's
initial surgery, the pediatric heart surgeon who'd been scheduled to
perform the operation died in an accident. The surgeon then doing
adults, who had also done pediatric surgery, took over the late surgeon's cases and did the repair. But the baby, instead of recovering,
lingered in the ICU, getting sicker and sicker. Because no one knew
what was wrong, she was recathed. Her left lung collapsed, and she
developed pneumonia; her whole body stayed swollen. Her doctors
feared kidney failure, so they put in a peritoneal catheter to dialyze
her. It seemed to Nina that the whole surgical and clinical team was
in that room meeting about her daughter every single day. Her cardiologist said he wouldn't rest until he figured out what was the
matter with her. The surgeon, though, told Nina that her daughter's
inability to recover might be due to her being overweight. "I was
told everything, from her being too fat—several things I was told,"
Nina said, shaking her head. Now that she was here and had finally
learned the actual reasons for S.'s difficulty in recovering, she was
angry. I repeated her comment: "The surgeon told you she wasn't
recovering because she was overweight?" "Yes, that's right," she said.
The cardiologist ultimately discovered the double arch squeezing
the airway, and that was when he knew she had to get to Cleveland.
This cardiologist had trained at the clinic and knew Dr. Mee. As
soon as the decision had been made to send her, the doctors at the

referring hospital no longer rounded in her room, Nina said. It was as if S. didn't exist anymore. The family made the trip to Cleveland via air ambulance in the middle of the night, S. by now in heart failure nearly two months after the original repair. On top of feeling terrified for her daughter, exhausted, and alone—as parents almost invariably did when their kids were here—Nina had begun to resent the referring hospital, though not her cardiologist. (In a subsequent telephone call, this cardiologist confirmed S.'s course, denied absolutely that the staff had ever stopped rounding on her, expressed regret over the surgeon's suggestion about her being overweight, and said, summarizing the situation, "It was felt that a second attempt [at surgical repair] would be too difficult, and we wanted the best possible person for this patient. . . . Other centers might try and try again, and they'd wind up with a dead patient on their hands and nobody'd know about it. We did the right thing.")

What made this situation unique was that the surgeon who'd done the unfortunate job was planning to visit the clinic the following week. He would arrive, by coincidence, while S. was still in the PICU.

The surgeon had traveled here to observe Dr. Mee and his staff. He'd brought his perfusionist along and was especially interested in watching the team's perfusion technique, he told me. His patients— his center performed about a hundred congenital cases a year— tended to be more edematous, or bloated, than normal, pointing to a perfusion problem. On the first day of the surgeon's visit, I did my best during rounds to stick close to Mee. Fairly early on, while the scrum was between rooms in the PICU corridor, the visitor asked about S. Mee told him she was here and doing all right. The surgeon then asked what had been wrong with his repair. Mee responded quietly but peevishly, explaining it to him exactly as he had explained it to me: "Patches don't turn corners," he said. He then went into some detail, using the curve between his thumb and index finger to represent the branching pulmonary artery, and also mentioned the loose patch. The visiting surgeon nodded seriously, even solicitously, I thought. And that was that—the scrum was on to the next room, the next presentation.

S.'s room was in the center of the corridor across from the main desk, and when the pack rounded on it, the visiting surgeon hung back and leaned on the desk. I hung back, too, curious. I was wearing a badge that said I was an ambassador (the honorific given to hospital volunteers), and the surgeon noticed it and asked why I was here. For a book on congenital heart surgery, I said. He asked if it was a medical text, and I replied no; then he asked if I'd written other books. When there was a convenient pause, I questioned him about S. The surgeon said that cardiology had failed to diagnose a double aortic arch and that this failure had caused the girl's problems. He made no mention of his own faulty work.

I wanted to ask more, wanted to say, "What about your repair?" and "Did you really tell the mom her daughter wasn't getting better because she was overweight?" But I didn't. It just wasn't done: "We're a closed guild," Mee once answered when I asked him why more people didn't know about the varying success rates among centers. It was impolitic to make a stink about what everyone already knew. I had heard this surgeon invite Mee's comments on his repair, and heard Mee respond. I'd had the benefit of Mee's description immediately after the surgery and could confirm the exact details of his findings in the operative note. I had checked, too, with Fackelmann, who'd also told me that patches didn't turn corners. When asked for further evaluation of the repair, Fackelmann had simply said, "It was terrible."

Obviously it was terrible: S. had lingered in the ICU, going downhill daily, and had nearly died from an operation that typically had less than a 2 percent mortality rate. The referring cardiologist would later tell Nina, she told me, that if they hadn't gotten S. to Cleveland when they did, she'd have died. The fact was, this kind of thing happened not infrequently; given the difficulty of congenital heart surgery and the near-impossibility of monitoring it in any meaningful way, situations like S.'s were inevitable. This, like much else in the healing arts, seemed to border on the primitive, which was ironic considering how high-tech we tended to think modern medicine and hospitals in America were. When things got really serious, life and death often came down not to those high-tech capabilities but rather to a few doctors' and nurses' skills, expertise, and intelligence—and just as in every other profession, as I'm not the first to note, a few were excellent, many were good, most were average, and some were mediocre. Given the stakes involved—the lives of children—we could *wish* that it were otherwise, but how could we *expect* it to be so when it was without exception this way in every other earthly endeavor, part of the natural spectrum of humanity?

This feeling nagged at me continually as I tried to navigate my way through this world, and it became acute when I saw what parents such as Shawna and Edward Cundiff, Nina, Angie and Bart H., and others had to go through. I kept hearing the nurse's best advice: *Don't get born with a heart defect.* And the New York surgeon's words to an audience of his colleagues: "I may get in trouble for saying this, but so what? In my opinion, the biggest risk factor is being admitted to the wrong institution."

The notion that something was wrong here stuck in my mind like a hook on the second day the visiting surgeon joined rounds. The first day Nina had not been in the room, but the second day she was, and I happened to be standing with her talking when she caught the surgeon's eye—the surgeon who had operated on her daughter, the surgeon whom she'd gotten to know during her daughter's many weeks in his ICU, the surgeon who had told her that S. wasn't getting better because she was too fat—and gave him a small, tentative wave. He nodded once in return, expressionless, then fell in with the pack of rounding doctors as they moved on to the next room. He was at the clinic for several days. Nina would tell me afterward that he'd never come to the room, never asked after S., never said hello. I supposed it was neither here nor there, but *I* would have expected at least a hello, along with some expression of concern. It had nothing to do with patient care at this point, of course. I had no idea how I'd react in a situation like that if I were the surgeon; it was a brutal one. Maybe anyone who did this work needed that kind of ego or denial, if that was what it was that prevented this man from speaking with Nina (and not just sheer mortification at what he'd done). It was impossibly unfair of me to judge this surgeon. But at the end of the day, thinking about it to myself, I couldn't help hearing Fackelmann's voice: *What an asshole.*

When it became clear to me that I wanted to write about S., I called the cardiologist and then the surgeon. Both returned my calls immediately and spoke frankly about the case. The surgeon defended his remark about the patient's being overweight: "Heavy kids have more problems. . . . She was a large girl, and with a thick chest wall it is harder to breathe and move the lung. . . . Her perfusion was quite good, but she was retaining CO_2.

"Our cardiologists missed the hypoplastic septum," he once again stressed, "and so we attached the patch to a flimsy muscle, and they missed the double arch." When I asked about the patch obstruction in the pulmonary artery, he said, "I do not remember. Perhaps Dr. Mee could tell you that." I mentioned the perhaps irrelevant but to me not insignificant fact that he had never said hello to Nina when he visited Cleveland. He denied it: "I did go to see her. I shook her hand. I don't remember what I said. Mom was not sociable." I relayed Nina's feeling that he had shown little concern for her daughter in Cleveland. "That is not true at all," the surgeon said.

He freely admitted, he said, that "I am not at the level of Dr. Mee. There are maybe five people in the world [who are that good]. You can't get to Dr. Mee's level when you have to do adult and other thoracic, when you only do a hundred [congenital] a year."

I traveled to Boston to have a look at Children's Hospital, the historical seat of congenital heart surgery when Robert Gross plied his trade, the place that had ushered in the current era under Aldo Castañeda, possibly the largest center in the world treating congenital heart defects. Children's had an ICU with twenty-four beds specifically for cardiac patients. Its four heart surgeons did more than a thousand cases a year among them, and the cardiology staff numbered thirty; the interventional cardiology staff, led by James Lock, one of the most renowned in his field, performed fourteen hundred catheterizations a year. The cardiovascular surgeon in chief, Richard Jonas, introduced me to the most junior partner, a surgeon named Joseph Forbess. Forbess had a crisp, straightforward appearance in scrubs and lab coat, his dark hair cut short. Born in Mobile, Alabama, and raised in northern Florida, he was thirty-seven years old, married, and the father of two daughters. Joe had wanted to be a doctor since he was a child. He'd attended Harvard Medical School, discovered he didn't like doing general surgery but found the physiology of cardiac defects fascinating, done his cardiac residency and a fellowship at Duke University, and ultimately proven himself a deft surgeon on animals during another fellowship in Boston under Jonas, who'd offered him a job as a heart surgeon.

"This is the most challenging physiology you can try to figure out and try to fix," Joe said. He'd been doing his own cases, 327 of them, for eleven months when I met him.

I'd wanted to speak with him because I was curious to know how a young surgeon started doing Norwoods on neonates, and how he did his first switch—those two very complex operations where skill made a huge difference in outcome. It seemed to me someone could learn to do this work only at a place like Boston, a hospital with a high volume of cases being performed by experienced surgeons who could teach, and an experienced team that included excellent anesthesiologists, perfusionists, and intensivists. How else to prevent mediocrity? I caught Joe outside the O.R., where he'd just finished a high-risk switch on a baby with transposition and a single coronary artery; it had gone well, he said. I told him why I wanted to talk with him. He nodded and said this was the only way to train—by performing those first high-risk operations alongside a senior surgeon (in his case John Mayer) before doing them alone.

"I can't imagine learning any other way," he said. "A mediocre surgeon like me just nestles right in here." (Joe would also call himself simply "a technician.") "I've got an unusual job," he continued. "I wake up in the morning knowing I'm gonna be pretty nervous all day." And then, "It's the system. I'm not Neil Armstrong. I'm just a guy, and I'm put in this system, and I can do switches. I've done thirty-five or something and I haven't had a death." I asked him how many Norwoods he'd done. "Fifty," he answered.

"How many deaths?"

"Three."

"Do you know why?" I asked.

"One had a small ascending aorta," he said. "One had bad veins. And one was probably a technical result." He'd had to redo the patch, he explained, extending the operation; the patient had had a series of postoperative problems possibly related to the long pump run, including failing kidneys, and finally died from sepsis. More than one account of the surgical personality had noted that those who readily blamed themselves for mistakes rather than shifting the blame elsewhere were invariably the best surgeons—though such admissions must come a little easier to someone whose results were as phenomenal as Forbess's.

Forbess had just described a 6 percent mortality with Norwoods, an astonishing figure that was later confirmed by Dr. Jonas. Jonas further noted that many of Forbess's cases had been complex Norwoods, babies with either a very low birth weight or other, associated problems. Mee's mortality, by way of comparison, was 16 percent, though he took on a high percentage of FUBARs. A mortality of 20 percent was considered standard at the best centers; the national average was more likely to be in the 40 percent range.* Forbess explained that he got so many hypoplasts because, as the junior surgeon, he took call most of the time and did all the cases that came in then. He tended to do fewer referrals, he said.

Forbess was clearly star material. His results were almost *too* good to be believable, even taking into account the extraordinary support system within Children's. Thirty-five switches without a death, performed by a surgeon in his first year of practice, was a phenomenal success rate by anyone's measure. (At the two smallest

---

*Howard P. Gutgesell, M.D., and Jennifer Gibson, M.S., "Management of Hypoplastic Left Heart Syndrome in the 1990s," *American Journal of Cardiology* 89 (April 1, 2002); Richard G. Ohye, M.D., and Edward L. Bove, M.D., "Current Status of Operations for Hypoplastic Left Heart Syndrome," *American College of Cardiology Current Journal Review,* January–February 2001.

centers in Ohio, by comparison, the total of seven switches performed by older surgeons had resulted in two deaths; the two largest centers, Cleveland and Cincinnati, had done seventeen and eleven switches, respectively, with no deaths.)

Back in his office Forbess told me that he went over the steps of each case in his mind on Monday evenings, during the cath conference, and if a case was particularly tricky, he might write up a flow sheet the night before the operation, sometimes going so far as to copy it onto the dry-erase board in the O.R. so everyone would know the exact plan, step by step. He called it his in-flight checklist.

"It's nerve-racking," he said of the work. "If you lose a kid— and I've lost five or six—it drives you crazy. It eats away at you." He could eventually get over the patients who died because they'd been "dealt a bad hand," as he put it—a single-ventricle heart with obstructed veins, say, or a tiny ascending aorta. "But if it's a technical error that results in a bad outcome, it's very difficult, because I'm a young parent myself, and these parents are so vested. I can't tell you how many times these parents have done in vitro, or they had a death of a spouse and they got remarried, and all the marbles are on this kid. The parents have told me, 'If this kid dies, it's gonna be . . .'—all the hopes and dreams of their entire assembled family are focused right on a little three-kilogram baby."

He explained how it was that he'd come to do so many Norwoods—or Stage Ones, as they're referred to here—and relatively fewer switches: many hospitals in the New England area did the arterial-switch operation, he said, but hardly any did the Norwood procedure. When he said this, I thought of the New York surgeon who'd quoted figures at a congenital heart congress, showing that in his state twelve different hospitals offered the switch operation, with mortality varying from 3 percent to 50 percent. I mentioned this to Forbess.

"Correct," he said immediately. "Which is a crime. I hope in your book you're going to get across to people that there is, I don't want to call it a crime or anything, but there is something being perpetrated on the American public."

What he meant, of course, was what I'd already seen and heard and thought about in various forms: this was nothing like general surgery; the difference between excellent and average here could be the difference between life and death; and parents were the least likely to know, or be capable of judging, whether a center was excellent or merely average.

"Stage Ones are the litmus test for a center," Forbess said. "There are a lot of places that do everything but Stage Ones. They're not like switches, where you'll have a fifteen percent mortality if you aren't up to speed. You'll have a hundred percent mortality—they'll *all* die."

When I called him later to go over what he'd said, he asked to clarify his remarks. An effort needed to be made in the public-health arena, he said, to address the huge discrepancies in outcomes among hospitals. And that effort, he said, "has to come from the government; it's not going to come from the hospitals." In any case, it was a problem that couldn't be ignored any longer: "It's the eight-hundred-pound gorilla of this profession."

Chuck Fraser, now the chief cardiac surgeon at Texas Children's, also commented, with the same animation as Forbess, on the wide variation in mortality in *his* state. "There are hospitals in Texas with a fifteen percent mortality," he said, adding, "and the general public doesn't realize it."

It was Jan Quaegebeur, a professor of surgery at Columbia Presbyterian and a ranking member, with Mee and a very few others, of the peds heart surgery pantheon, who had presented the figures from New York, numbers compiled by that state's Department of Health. With a mandatory reporting policy for all cardiac interventions, New York was one of the few states rigorously organizing statistics on congenital heart surgery and making them available to the public. Quaegebeur, who worked with the Department of Health's cardiac services program, said that it was difficult to evaluate surgical outcomes in congenital because of the small numbers relative to surgeries for acquired heart disease, as well as the infinite variability of the defects themselves, which made all but the most straightforward types incomparable. I called him after his talk, and he reiterated, "Many people are unaware of the differences there are between centers, that there are centers of excellence and centers that do the best they can.

"It's not easy to get this data to the public," he concluded. "Some centers are less interested in having outcomes published in the newspaper. . . . Patterns of referral are established, and institutions compete with each other."

When data that reflected negatively on physicians *did* reach the public, it was typically in the context of heated media coverage on centers where there were suspicions of unnecessary deaths, whispers about obnoxiously arrogant surgeons, or messy internal politics—stories often leaked to investigative reporters. Such scandals could result in sweeping changes in a center, as happened at the Bristol Royal Infirmary in England. In the summer of 2001, responding to accusations made by a "whistle-blower" who had leaked word of bad results to the media, the government released the findings of a lengthy and expensive investigation it had undertaken. According to the BBC, the report concluded that "between thirty and thirty-five children who underwent open-heart surgery at the Bristol Royal Infirmary between 1991 and 1995 died unnecessarily as a result of substandard care." Two surgeons were barred from operating after investigators found that they had continued to perform surgery on children despite poor survival rates, and that the hospital's chief executive had covered up their results. A television documentary named the doctors involved and cited distinguished work in the field by such prominent surgeons as Bill Brawn and Marc de Leval.

Two other stories surfaced during my sojourn at the Cleveland Clinic. The first one broke with a report in the *Seattle Weekly*, sparked by a series of anonymous letters apparently sent by a member of Seattle's Children's staff to the parents of children who'd died there, implying that the hospital had been negligent. No wrongdoing was ever confirmed or proved, but the scrutiny did bring to light internal strife between the "difficult" chief surgeon and ICU staff and cardiologists. The surgeon ultimately left the hospital. That surgeon's partner, Brian Duncan, was later to replace Jonathan Drummond-Webb in Cleveland. (Duncan, forty-two years old, while not denying the internal strife, would say that the events in Seattle had had little to do with his move, which was instead prompted by the opportunity to work with someone of Roger Mee's caliber.)

The second story reported in the *Denver Post* was the more disturbing of the two. Cardiologists at that city's Children's Hospital regularly referred neonates with complex heart problems to other centers, but the hospital continued to do a small number of neonates. The cardiologists, however, believed in the veracity of the volume-outcome relationship (the more cases, the higher the success rate) and felt that all neonates should receive the best possible care. Thus among themselves they made it a policy to refer all neonates to centers in San Francisco, Boston, Cleveland, and elsewhere.

"There was a growing perception that results were less than we expected," said Mark Boucek, then head of cardiology there, speaking on behalf of the group. After this came to the attention of the administration, Boucek stepped down. Boucek, according to the *Denver Post*, said that he was asked to resign as chief because of these actions. Boucek's boss told the *Post* that the request that Boucek step down was, the *Post* wrote, "the culmination of 'broad and long-standing' misunderstandings."

Boucek confirmed the events by telephone, and while noting he was uncomfortable in this kind of spotlight, he did say that the hospital's claim that he was demoted because of long-standing problems strained credibility, given the timing of his enforced resignation and the fact that two years earlier he had been given a substantial raise in order to keep him at the hospital. He suggested that the hospital might have been trying to "rein in behavior" by "cutting off the head of the beast."

The hospital continues to deny this. Responding to me by e-mail, the hospital wrote, "Dr. Mark Boucek was not reprimanded for anything. We support the cardiologists' decision to refer as they see fit. And as always, it is up to the family where they choose to receive care.

"Dr. Boucek's stepping down," the hospital continued, "was part of long-standing differences involving the administration of the department, which included working relationships." The timing, the hospital noted, was "not necessarily a factor.

"Dr. Boucek," it concluded, "remains a prominent and well-respected practicing cardiologist on staff here at The Children's

Hospital." The hospital noted it initiated an external review in December 2000.

The result of Boucek's actions and the attention in the *Denver Post* almost certainly has been improved patient care at Denver Children's. Noting that poor surgical results are rarely the fault of a single surgeon, but rather the fault of several weak links in a system, Boucek said that the hospital had hired a "world-class" surgeon and created an ICU and an anesthesia unit devoted to cardiac patients. Boucek noted, "It takes a big event to change things," referring to his group's policy to refer all neonates out. Political decisions, he said, "are never in the best interests of the patient," and he noted the chilling message the hospital's actions sent: "You can imagine the future, cardiologists realizing they're not as free to make these decisions."

Mee likewise had told me, "Cardiologists aren't always responsible for where they send their patients," adding that "cardiologists should be able to question surgical results. . . . A surgeon has to respect the cardiologist's decisions."

Likewise, in Bristol, a new surgical staff and new procedures, instituted following the scandal, have resulted in mortality half that of the national average. Public scrutiny is difficult for those involved—most hospitals still don't make their results public—but it seems unlikely to make a hospital worse and typically benefits patients.

For me, the Denver story was every bit as scary as reports about surgeons who continue to operate in the face of consistently poor results. Here was a group of cardiologists, after all, who had tried to do what was best for their patients by sending them to the centers they thought best for the job, and they were apparently reprimanded for it. The flip side of this scariness was the implication that what this group had done was unusual, which meant that most cardiologists refer patients within their systems regardless of those systems' results or face professional risk. Surgery is, finally, the moneymaker.

And yet this kind of reciprocity was all but inevitable except in the most complex cases. Using a number published by the American Heart Association in its statistical update—25,000 open-heart surgeries performed each year on children—and an upper limit per heart surgeon of, let's say, 250 operations a year, it would take a hundred surgeons to meet the demand for open-heart procedures alone. And of those one hundred, how many were of Roger Mee's, Ed Bove's, Jan Quaegebeur's caliber? Ten? Twenty? If those twenty did their 250 each—say, the most complex of all cases—that would *still* leave 20,000 more operations to be done by lesser surgeons. Who would willingly choose to hand his or her child over to one of these?

The system was far more feudal than that, though, territorial and driven, like any other business, by financial pressures and long-established patterns. Furthermore, no one charged that anyone was out deliberately to hurt patients or do anything other than treat them. Even the Bristol report concluded by describing itself as "an account of people who cared greatly about human suffering, and were dedicated and well-motivated." Unfortunately, their behavior was flawed, and "many failed to communicate with each other, and to work effectively for the interests of their patients. There was a lack of leadership, and of teamwork."

Now and then while writing this book, I asked an old friend, an associate professor and attending physician at the University of Virginia Health System, for his perspective on difficult questions. In this instance he wrote back, "Most doctors who hurt people do so while trying sincerely to help."

But the possibility that some doctors might consciously not do what was best for their patients continued to bother me. And whenever I probed in the direction of those benign-sounding "referral patterns," people became guarded. In more casual moments Mee might seem merely peeved. I would ask something like, "How can a cardiologist in Columbus, say, not refer a switch to you?"

"Columbus? How about five minutes down the street?" The reference was to Rainbow Babies' and Children's Hospital, part of University Hospitals, archrival of the clinic. When I pressed the point, though, even Mee grew cautionary. "Don't be judgmental," he warned.

"Is it unreasonable to ask these questions?" I responded.

"It's reasonable to ask how decisions are made," he said.

Dan Murphy, who was always a good, if careful, source of explanations and information, could tell me that here referrals were easy for him, because he knew he was referring to one of the best. But Murphy was soon to leave the Cleveland Clinic for Stanford. When he accepted the job offer, he didn't know who the surgeon would be at Stanford; he and his wife simply wanted to relocate to their home state. When it turned out that his new employer had recruited Frank Hanley, an acknowledged leader in congenital heart surgery, Murphy was relieved. "It makes my job a lot easier," he told me by phone from California. Didn't that imply, though, that while it might've been harder for him to refer patients to a lesser surgeon, he would have done it anyway? It meant, he replied, that the decision-making process would've been a lot more complex and difficult if he'd had to refer to a surgeon younger and less experienced than a Hanley or a Mee.

Unlike most I spoke to, who deflected the toughest questions, one doctor at the clinic went out of her way to address this issue. Interventional cardiologist Tamar Preminger, a lanky thirty-nine-year-old from Manhattan with straight dark hair and a square jaw, not only was willing to talk about "referral patterns" but had the experience to support her convictions, having worked at Children's in Boston, Children's Hospital of Philadelphia (CHOP), and Great Ormond Street in London, three of the world's best. She had also spent five and a half years at the rival Rainbow Hospital before being hired away by the Cleveland Clinic. I couldn't expect her to be objective regarding her former employer because she disapproved of it so strongly ("I used to send Roger cases when I was at Rainbow," she said, "and I got reprimanded"), but she was one of the few who would speak openly about the general situation. When I asked Mee about her, he raised his eyebrows and said, "Tamar is reckless politically. But I admire her more than anyone in the department for it."

In the late spring of 2001, the chief surgeon at Rainbow, Phil Smith, jumped ship with many others on the congenital staff to open a center at Akron Children's Hospital, thirty miles south of Cleveland. This left Rainbow Babies' and Children's Hospital with no pediatric heart surgeon. I happened to be hanging around Mee's office one afternoon when Preminger wandered by in her customary green scrubs, lab coat, and red clogs. Mee, on his way out of the office for the day, had stopped to look over some e-mails and glance at a six-inch-high stack of referrals that Debbie had left for him on the shelf-partition above her desk.

Preminger and Mee exchanged greetings, briefly discussed a patient, and then Preminger asked, "Have you heard what's going on at Rainbow?"

"No," Mee said.

"Apparently they're bringing up a surgeon from Columbus one day a week."

"Really?" Mee said. As he said it, he let his knees buckle forward. "Who?"

"I don't know," she said. "I just hope nothing goes wrong. They're two and a half hours away." She shook her head and explained to me how impractical this was as Mee nodded in agreement. If, for instance, a postop patient arrested or there was excessive bleeding, who would reopen the chest with the surgeon 150 miles distant? Tamar said, "The only ones who don't know are the parents."

"Right," Mee said. "Keep the parents in the dark."

I had spoken to Preminger on several previous occasions about these very issues, which infuriated her. "I'll never stop being idealistic," she'd said.

"Those poor parents who give their healthy three-year-old over with a PDA, and then to have them lose the coil!" she exclaimed to both of us. She'd heard not long before of a mishap at another hospital, where a catheterization to repair a patent ductus arteriosus by plugging it up with a coiled wire had gone awry: the cardiologist had let go of the coil, which had then been accidentally pumped elsewhere in the patient's vasculature.

"And then tell the parents," Mee said, his tone turning to one of mock gravity, "what a *difficult* procedure it was, and how they *saved the child's life*."

I asked them why Rainbow would bring in a freelance surgeon, and Mee said, "They just want to keep it ticking." It was very difficult, he said, to restart a heart center after it'd shut down.

Tamar said, "Their decision-making process has always left me . . . breathless."

Later I asked her why Rainbow hadn't taken the upheaval as an opportunity to join forces with the clinic. Mee had told me that when he first got to Cleveland, his partner had approached Rainbow about creating a combined center—the more heart centers in a given area, he believed, the worse the quality of care. Rainbow had declined, however, to enter into any kind of partnership. Now seemed like the perfect time for it to revisit the idea: cardiology could still bring in the lion's share of profits for both places, and the patients would get the benefit of both services.

But Tamar said, "They knew we'd be open to it [a partnership], but they made a contract with the center with the worst statistics in the state."

I called Rainbow Babies' and Children's Hospital to ask for a response to these issues. By informal reputation, Rainbow is the place for kids in Cleveland, the community's hospital, while the clinic is the forbidding "evil empire," as it's sometimes referred to at Rainbow according to doctors who have worked at both places—a cold behemoth that caters to wealthy Arabs and appears to care more about VIP patients than the huddled masses of Cleveland. Renowned in many areas, the 226-bed Rainbow was ranked seventh in the country in pediatrics by *U.S. News & World Report* in 2002. It employs a staff of 850 pediatric specialists and has one of the most respected neonatology units in the country; its NICU treats 1,200 patients a year, and its PICU 1,600 patients a year. At the time I called, Rainbow had just hired a new surgeon and was on course to resume its 200-plus caseload of pediatric heart surgeries.

I was invited to speak with Kenneth Zahka, division chief of pediatric cardiology (and Preminger's former boss), and Meri Armour, senior vice president of women's and children's services for University Hospitals of Cleveland. We sat in Dr. Zahka's office along with Eileen Korey, director of news services for University Hospitals, who asked to tape-record our conversation. She pointed her tape recorder at me, and I pointed mine at Dr. Zahka. Nothing that surprised me was said: the hospital did not reprimand cardiologists for referring out, both Zahka and Armour said; they defended their decision to use the team from Columbus; they do sometimes refer to Roger Mee, as they do to Ed Bove or Frank Hanley or whoever they believe is the best match for the patient, given the family's desires.

"I would say that most of the kids who came to us who needed complex surgery went to the Cleveland Clinic . . ." said Zahka, referring to the period after the departure of their previous surgeon. "You don't feel like you're going to be criticized if you send somebody out."

He added, "I have to wake up in the morning; I have to look in the mirror. One thing that I hope I have at the end of the day is my integrity, and I think I have it."

Armour said, "The faculty here practice independently, they really have their own practice plans, they really are their own providers, and one of the things that's special about Rainbow is that this decision doesn't just rest in Ken's hands. . . . There are cath rounds, we have decisions made by a team, and people look at [each case] and say, 'What's the best thing to do for this kid?' "

Just to be clear, I reiterated the question, Would a doctor be reprimanded for referring out? Zahka made a face at me and said, "Oh, my goodness. Goodness gracious." Armour shook her head at me: definitely not.

Regarding their choice of Columbus, Zahka said, "They were nearby; they were willing to come and be supportive of another children's hospital. And they had very great clarity about what they were willing to do—they were going to bring two surgeons, their own perfusionist; they were going to stay after the patient was done; and they had very clear guidelines about the types of patients they were going to do. And they were exactly in concert with what we do. We weren't going to do neonates—we were going to do ASDs, VSDs, simple tetralogy, coarctation in an older child. . . . Those are operations which everybody in the state has a hundred percent success with. And keep in mind we have a NICU, a PICU, anesthesia, cardiology; we have a mature team here. I had every confidence that they would do a great job. . . . It was actually a wonderful relationship."

Regarding the question of a joint program, Zahka said, "We would be happy to do that. But every time it actually came up and somebody from UH started to make contact, we were rebuffed."

Armour added, "Remember that the clinic is a closed medical staff"—that is, it's a group practice, not a private practice. "If we had a neonate here that needed to be operated on, we had to transfer them over there. It's always going to be a one-way street; it can't be a collaboration. . . . They wouldn't have privileges to operate at another institution."

I asked to sit down with Preminger in her office to talk more about these issues, about how cardiologists decided where to send their patients for surgery. As always, she was straightforward.

"It's a very personal thing," she said of this specialty. "Pediatric cardiology is a very small field. People know everybody, they know everybody's results, so there's no excuse to say, 'Oh, I didn't know this person was here.' What should happen, ideally? We should send patients to the best person to do [their surgery]. And everyone in this field knows that the more you do, the better you are. Does this happen? No, because most people aren't motivated by that, by doing what's best for the patient. That's what you're supposed to do as a physician.

"But you need someone who—this sounds so terrible—who has some morals and values in order to do this. And parents have no idea," she said. "That's the most criminal thing. Parents of course want what's best for their children and they have no idea.

"What you have to say to people is that Rainbow's terrific for some things, but the best heart surgeon in the world, one of the best in the world, is at the Cleveland Clinic. When I approach cardiologists, they'll say, 'I decide based on the gravity of the lesion; complex I send to Roger.' My response to that would be if my child had an ASD"—a defect that can be easily repaired, with zero expected mortality—"there's a tiny chance that something can go wrong. But if it does, I'd want to know I did the best for that child. So to me it doesn't matter if it's simple or not simple; it's still heart surgery, and things can still happen.

"This is someone's *child*. There's nothing more precious than a child. You do the right thing."

Evident though her antipathy toward Rainbow's heart department was, Preminger made a point of underscoring that her observations were not specific to a single institution; this sort of thing went on in many places, across many medical specialties. She'd wanted to give me her thoughts on the subject because, she said, she knew that few other doctors would go on record about this, and she felt compelled to do all she could to call attention to what she believed was an important issue. "It's huge," she said. "It's huge. And it's not just in this field. And I guess I can say this because I'm a physician: not all physicians do the right thing."

Was Preminger holding her colleagues to impossible standards? Was she oversimplifying a complex situation? Maybe. Did most physicians do what they thought was best for their patients? Probably. But what was undeniable, and verifiable, was that though some centers had better results than others, cardiologists continued to refer their patients to the lesser places, a fact that supports Preminger's claims. It seemed to me that some kind of official monitoring and publication of surgical results would be useful to the public. It might not help doctors, but it seemed unlikely to harm patients, and it would probably make patient care better overall by forcing heart centers to be accountable to the community for their results. During the strife in Denver the hospital there hired outside doctors to evaluate its practices so it could improve them. Mee sometimes visited centers to do this type of evaluation when they were having trouble; he'd found that when results were bad, the problem typically lay not in a single surgeon or doctor but rather in systemwide deficiencies. In New York State similar monitoring had led to one center's voluntary closing of its congenital heart-surgery program because of poor surgical results.

In the state of Ohio a group within the Department of Health—the Bureau for Children with Medical Handicaps, whose mission was to ensure that all children with serious medical problems got decent treatment regardless of their insurance status or ability to pay—collected data on pediatric cardiac interventions performed by the four centers that offered them: the Cleveland Clinic, Rainbow Babies' and Children's, Columbus Children's, and Children's Hospital Medical Center in Cincinnati. (Beginning in 2001 Akron Children's Hospital would be added to the list.)

From this data I would learn that for open-heart surgery in the previous year, for instance, the Cleveland Clinic had reported an overall mortality of 2.95 percent, Rainbow 2.5 percent, Columbus 7.79 percent, and Cincinnati 6.15 percent; that neonatal mortality had been, respectively, 8.82, 16.67, 27.27, and 21.05 percent; that the clinic had done eleven Norwoods and lost two of those patients (18 percent), that Rainbow had done five and lost none, that Columbus had done seven and lost five (71 percent), and that Cincinnati had done eleven and lost four (36 percent). But even these statistics could be manipulated. I could truthfully assert, for example, that Rainbow had had a 25 percent mortality for the arterial-switch operation, and Columbus a 33 percent rate, while the other centers had reported zero mortality—and all for a procedure in which the biggest centers routinely expected a mortality of 3 percent or lower. But Rainbow had done only four switches in total, and Columbus just three—so did one death apiece really *mean* anything, given such small numbers? Were these figures helpful? Rainbow had reported a better overall mortality than the Cleveland Clinic that year—what did *that* mean? Was it a better heart center? If my child needed a Norwood, I wouldn't take him or her to Columbus based on the data—that seemed a safe conclusion. And yet some cardiologists continued to refer patients there for that complicated surgery. (The following year mortality would drop to 33 percent for this procedure, below some national averages, but the year after that it would go up again, to 80 percent of five patients, according to BCMH data. Subsequently, and of its own accord, Columbus Children's Hospital would take substantial measures to improve its heart center, announcing in May 2002 the hiring of a new chief of pediatric heart surgery, a cardiologist specializing in cardiac critical care, and an additional interventional cardiologist.)

At the Children's Hospital in Denver all cardiologists had

begun to refer out the neonatal open-heart surgeries because the mortality at their own center was unacceptably high, but there was no law requiring them to do this, and no outside body even suggesting it; they apparently did it on their own accord, and not without professional risk. The entire specialty, like most areas of medicine, was exclusively self-monitored (where it was monitored at all, that is), and certainly the issues were not open to public debate. There was no justice in this world, and luck could make all the difference. When it came to repairing babies' hearts, some places were good, and others "did the best they could."

Many in this field favored an idea known as regionalism, a national system whereby several large centers around the country would be designated and permitted to perform and teach congenital heart surgery; several large centers doing a high volume, studies suggest, would likely result in more live children. This idea seems so obvious as to make it astonishing that we don't have such a system in place to best serve infants and children. All data point to the unsurprising idea that the more surgery a center performs, the better its results will likely be. Numerous studies show that centers performing fewer operations tend to have a substantially higher mortality than centers that do a lot of operations, and the difference is especially acute in cases involving complex defects and very young babies. No data suggest that small centers, as a rule, do better.

When Kathy Jenkins, a forty-one-year-old cardiologist at Boston Children's, entered her specialty, she was amazed by the hugely varying mortality rates at various centers, so she set out to study the subject. "The striking finding," she says, "is that it is clear that there is a volume-outcome relationship"—higher volume, better outcomes (fewer deaths). But it's not as simple as that, because there are some small centers with stellar results and some high-volume places with bad results. It's not a smooth curve, it's not proportionate, and you can't institute a referral policy based on a center's volume alone; volume is thus not the sole factor, but is in fact one main factor among several. "Little is known about what's driving the volume-outcome relationship," she says. The volume-outcome question is not limited to the peds heart surgery specialty; it's a factor in many specialties. (According to one study, there does not appear to be a volume-mortality relationship for coronary artery bypass surgery, for instance, but there is such a relationship for complex cancer surgery.) Its effect regarding complex congenital heart surgery is not insubstantial, given, Jenkins says, that "there is a very serious variability in mortality among centers; it's real, and it applies to children going into the system."

What do these facts and substantial support data result in? Zippo. Ohio is a representative example. According to the volume-outcome studies, the citizens of that state would be better served by a single center the size of the Boston Children's program, which does about a thousand cases a year, or if not one center then two, performing the one thousand pediatric heart surgeries that take place in the state each year. Instead, until spring 2001, it had four centers. And then what happened? Unhappy surgeons and cardiologists left Rainbow Babies' and Children's and joined Akron Children's Hospital, thirty miles to the south, which was, I'd wager, eager to add this lucrative specialty to its repertoire of services, rather than refer them up to Cleveland as it had been doing. Thus, Ohio now has five centers that offer heart surgery on babies and children. So instead of fewer centers doing more cases (likely with a better statewide mortality rate), it has more centers doing fewer cases—Roger Mee and the clinic included, since Akron cardiologists no longer need to refer cases to him, as they had been doing. While there's no way of knowing this for certain, it's unlikely that any of these events resulted in better patient care, if the studies are accurate. In any case, the events do not seem to have been motivated by a desire to improve outcomes in the state.

When Mee started out, the so-called learning curve was an accepted fact, risk was high, and deaths were commonplace in pediatric cardiac surgery. But now there is no longer any room for a learning curve—the big centers and many small centers know how to do it and do it well. Consumers are becoming smarter, and doctors are less likely to be deified. All of which is in the patient's best interests, but it still seemed harrowingly hit-or-miss to me.

10. The Norwood

Telephones ring continually in the O.R.—so often, in fact, that their gentle staccato signal is noticed by almost no one. Typically the circulating nurse will lift the receiver of a wall phone by the computer terminal and update the cardiac nurse on how the procedure is going. Occasionally one of the beepers parked by the computer will go off, and she will respond on behalf of a doctor scrubbed in on the case. Mee usually leaves *his* beeper on the belt of his black slacks, draped over a chair in his office. But there's also a phone attached to the anesthesiologist's monitor cart, so if the scrub nurse is occupied, Julie or Paula or Emad can answer, as Julie did on Valentine's Day.

The O.R. was quiet but for the usual hum of the heart-lung machine. Mee was deep into a difficult repair on a boy from Kuwait with a hypoplastic right ventricle (*hypo,* meaning "under," and *plastic,* rooted in the Greek verb meaning "to form" or "to mold"— thus an underformed right ventricle) and critical pulmonary stenosis, a narrowing of the pulmonary tract through which blood flows from the right ventricle to the lungs. Today's case would be another instance of Mee's attempting to jury-rig a biventricular repair, as he'd done for Patrick Cundiff the day before. Now that he was inside the chest, he had to make his decision, so he stared at the heart, felt it, thought about the echo (which showed good right ventricular function), and speculated as to whether this right ventricle could

do the work of pumping blood to the lungs. Or should he, alterna-
tively, start the kid down the single-ventricle road by routing blood
directly from the vena cava into the pulmonary artery in a procedure called a Glenn, to be followed a few years later by a Fontan,
thereby bypassing the right ventricle altogether? He inserted a needle in the pulmonary artery and registered a pressure there of
eleven millimeters of mercury, low enough for the Glenn. But after
having a look around inside the child's heart, he believed he could
fix or amend its defective physical structures—the bad ventricle,
the tricuspid valve, and the pulmonary valve—and he guessed that
the right ventricle could do a lifetime's worth of pumping, so he
went for the biventricular repair. It was a long, tense case. When he
was less than halfway through, the phone rang, and Julie, the anesthesiologist, answered.

She said into the mouthpiece, "Uh-huh, uh-huh," then lowered
the receiver but didn't hang up; she stepped close to the cage to address Mee.

"Roger," she said, "they've just delivered a hypoplast without
an ASD. They can't get sats above forty."

Mee paused in his work and looked up. He rarely took his gaze
off the field. He appeared to be almost angry, as if to say, *Look
where we are—what do you expect me to do?* His eyebrows rose, he
tipped his head slightly forward to regard Julie over his loupes and
said, "*Well.* Better get 'im to the cath lab." And he returned his
focus to the hypoplastic right ventricle from Kuwait.

The situation was urgent, but luckily the cath lab had just freed up.
Geoff Lane, sitting behind the glass partition of the control booth,
finished his dictation on the previous case and mumbled, "There is
no end to the paperwork." Two cath-lab nurses remained in the
room. "We're going to do the baby," one said to the other. "She
doesn't know its name." The other nurse, irritated as she began to
prep the room, said, "They live like pigs down here—trash up the
walls, blood from every patient all over the floor." Geoff Lane gathered his papers and left to have a look at the new case. Ironically,

the one he and staff cardiologist Lourdes Prieto had just finished was an ASD closure using a plug called a CardioSEAL. They would now have to *create* an ASD, open a hole between the atria, in an emergency procedure called a septostomy, on a critically ill baby just minutes old, to allow blood backing up in the lungs a way out. Prieto had just left and was now heading for the O.R. to speak with Dr. Mee.

She wore green scrubs. Her salt-and-pepper hair was in tight curls, and the features of her face—cheeks, nose, lips—were sharp, her eyes dark. Born in Cuba, where she'd lived till age fifteen, she retained a rich Hispanic accent. She opened the door to the O.R. and called to Mee, explaining the case.

He didn't look up, remained hunched over the open chest. He said, "What's the problem?"

Her head bobbed from one side to the other.

PRIETO: Septostomy in this situation has a ninety-five percent mortality.
MEE: Where?
PRIETO: Well, at this institution there has been one hundred percent mortality.
MEE: So what?
PRIETO: [speechless].

She might well have been thinking of the last such rare, urgent case she'd done with the chief cardiologist, when the catheter had gone into the left atrium and, she suspects, out the other side, and the baby had died. The one before that hadn't survived, either; in fact, she'd never known of a baby with this defect who lived.

Mee said, "When you balloon the kid, just dilate it, don't pull it."

Geoff Lane, who would assist Prieto in this procedure, walked the two flights up to the echo lab to have another look at the echocardiogram of the newborn's heart. Suzy Golz, a cardiac sonographer, stared at the video as Geoff sat beside her, watching the thirteen-inch screen.

"Is that aorta atretic?" Geoff asked.

"Almost," Suzy said, acknowledging that there was virtually no flow through this major vessel. "God, look at that ventricle—it's just a big piece of meat."

This was of course the classic feature of the heart defect known as hypoplastic left heart syndrome (HLHS): a thickened and useless left ventricle, ordinarily the main pump of the heart. With this defect, it doesn't fill and scarcely contracts. In most cases of HLHS blood returning from the lungs into the left atrium pushes its way through an ASD into the right atrium, where it dumps into the right ventricle, thence to be launched into the pulmonary artery; some of this blood goes through the ductus and up the aorta to feed the body. But in perhaps 5 percent of HLHS babies there is no ASD, so the blood has nowhere to go; it turns the left atrium into a taut balloon as blood backs up in the lungs. These babies get sick as stink and then usually die, fairly quickly. The only hope is to push a catheter into a vein in the baby's groin, maneuver it into the right atrium, and pop a hole in the septal wall, which has become thick and tough from the intense pressure.

Dan Murphy strolled into the echo lab and took a look, then confirmed that the aorta was atretic.

"But the arch is not bad," Suzy said.

"Coarct," Dan said.

"Yeah, but not bad."

Then they all focused on the left atrium, that enormous, blood-swollen balloon, and noted how difficult this one was going to be—especially pushing a needle through that wall of tissue without having it come out the other side, as had happened when Lourdes assisted the last time.

"It's kind of like putting another hole in your belt loop," Dan said. "On your leg. Using an awl."

Geoff headed up to the nursery, which was buzzing with activity almost to the point of being chaotic. The small, square room was packed with people, and the frantic energy seemed palpable among the residents and attendings, the neonatologist, transplant cardiologist Maryanne Kichuk, the respiratory therapists, echocardiologist

Adel Younoszai, the nurses. The only people not moving, the stillest people in the room, were the baby's parents. Brian Mangan was tall. Standing at the rolling warmer table where his seventh child lay, he seemed older, with thinning, close-cropped gray hair and watery red eyes. Patricia Mangan, his wife, who had delivered her breach baby less than an hour ago by cesarean section, sat in a wheelchair, touching her newborn boy. She had not yet gotten to hold him and had had to plead to be wheeled in here to, as she told the doctors, "say good-bye to the baby." She was crying, but not weeping freely; instead she seemed to be bleeding tears, exhausted and scared and sad.

Prieto had already told the Mangans that there was a 95 percent chance their baby wouldn't make it through the procedure—they should expect him to die, she said. The baby, Kenneth, looked normal aside from his purple extremities and generally purplish hue. But inside, even though his lungs were already stiff with backed-up blood, his healthy right ventricle kept pumping more in; until Prieto could pop a hole in the atrial septum, the blood had nowhere else to go. Furthermore, it was turning to vinegar from all this stress: because his tissues were getting little oxygen, his metabolism had changed from aerobic to anaerobic, one by-product of which was lactic acid. The acid level was a marker of how sick he was, and it also had an impact on his organ function. As Marc Harrison put it, "He's making a dead-on sprint for the finish." Severe organ damage, followed by death. Kenneth was hand-ventilated while the doctors struggled to get lines into his tiny vessels.

When they'd found arteries into which they could pump drugs, a team of four wheeled the baby out of the nursery to the elevator, past waiting families, down six floors, then through several corridors to the cath lab. In the hall outside the cath lab a nurse offered people Tootsie Rolls. Bowls of little heart-shaped Valentine's candy were everywhere.

As always, the cath lab reminded me of a space station. A glass wall separated the control booth from the floor, to protect staff from the X-ray radiation. The technician in the booth would type every step of every procedure, and the time it was performed, into a computer, and could view various screens giving different vantages of the angiograms. A microphone broadcast his voice to the dimly lighted room beyond the partition where several screens were set up for viewing and where an enormous white machine, the gigantic X-ray tube, was suspended over the catheterization table on a C arm to capture angiograms and do fluoroscopies. Of the 9,000 catheterizations done at the clinic each year, about 350 to 500 are done on children, whether for diagnostic purposes, to take biopsies (tiny bites of tissue) from transplanted hearts that can then be examined for signs of rejection, or, increasingly, to fix problems that once could be repaired only through open-heart surgery—ASDs and the occasional VSD, obstructed veins and stuck valves, pinched aortas.

Adel Younoszai came in to echo the baby so Prieto could get an ultrasonic view of the heart. (In angiography only the catheter itself is visible until radiopaque dye is sent into the heart, forcing the cardiologist to guess where he or she is.) While setting up, Adel told me, "This is a procedure that is almost universally fatal," though he added hopefully, "But babies are amazingly resilient." He didn't have the surgeon's mentality, preferring the low-stress intellectual work of echocardiography, and was happy not to be in Prieto's shoes today. Adel thanked Maryanne Kichuk for her help in getting the baby down here safely. Maryanne said, "They're so disorganized up there. They're well intentioned, but they aren't used to this."

When the phone rang, Adel was nearby and answered it. "Emad," he said, "it's you. The baby's being strapped down. He's very sick. . . . Yes, we have a ventilator." The baby was just over an hour old when Emad Mossad arrived to intubate him. Amy Toth, a cardiology nurse, began threading the clear plastic nasogastric tube into Kenneth's left nostril, but its placement went awry, and she pulled it out again. The baby writhed, his mouth wide open but silent, his face purple. "I'm sorry, honey," Amy said. She got it on the second try.

The mood in the room was tense but not panicked. People spoke casually to one another, as if content to be engaged together in a difficult but satisfying task. A cath-lab nurse announced, "His sats are dropping," but Emad said, "It's OK." Oxygen saturation was not what he was worried about now; acidosis was his main concern. Prieto, now scrubbed, agreed as she joined Emad at the table: "Don't even look at the sats." As the nurse swabbed the baby's groin with Betadine, she said, "It's Valentine's Day, and nobody gave me a card."

After she'd laid out all the tools and catheters and balloons and syringes she'd need on a setup table, the procedure began. Prieto inserted a catheter into the right femoral vein at the baby's groin and advanced it toward his heart. Geoff Lane said, "Let's have a little less talking," and the room quieted.

Prieto turned to Lane and asked, "Do you want to do a septostomy?"

"*No*," he said.

Prieto, in teaching mode, wanted to make sure that Lane knew the risks of jerking the balloon through, and clearly he did. Typically, interatrial holes are created by putting a catheter through the septum, inflating a little balloon, and then jerking it through, as Lane had done with Connor Kasnik's heart a couple of months earlier. But in this case the pressure in the left atrium had made the septal wall extremely tough, and if Prieto tried to yank a balloon through, she might rip open the heart and kill the baby. This was why Mee had advised, "Dilate it, don't pull it." Prieto's plan was to get a catheter into the right atrium, position it against the septal wall, and then push a needle through this wall, *gently*, so as to avoid going out the back, a distance of less than an inch. The needle she would use had an inflatable balloon over it; once she got the needle into the left atrium, she could inflate the balloon to open up a hole.

Amy read off another blood gas and noted a base excess of minus 11. In his exotic Egyptian accent, Emad said, "Yikes, he's getting worse." He'd already given the baby two doses of bicarb. When Amy read off the glucose level, Emad decided to give some albumen and asked, "Do we have any calcium?"

Prieto had now begun to advance the catheter into the vena cava, the main vessel ending at the right atrium. All could see the catheter move on the screen as it bent and reversed direction, having hit the top of the atrium. Because the heart itself remained invisible, much of this was guesswork. She asked, "Adel, can you tell me where the needle is?"

Adel, scrutinizing his own screen, could see the structures of the heart, unlike Prieto, and he could also see the catheter. "It looks like you're right up against the coronary sinus," he told her.

"We're going to start again," Prieto said, pulling back on the catheter. The baby had an unusually large coronary sinus, or opening—the vessel through which the coronary veins send blue blood directly back to the right atrium—and she kept getting stuck here. Standing straight upright at the table, Prieto worked the guide wires in her hands at table level, all the while staring at a screen. This was high-tech, high-stakes medicine, but her physical movements were not so very far from those of someone trying to jimmy open a car door using a coat hanger—she worked at that remove. When she pushed the wire forward and it buckled, she knew she was stuck in the coronary sinus again, and Adel confirmed it. "That's a huge coronary sinus," she said. She pulled back and tried again.

The two tools—echocardiogram and X-ray machine—combined effectively in this instance. Adel could help her visualize the atrial septum where it ballooned out into the right atrium, a gift of millimeters. Knowing the shape of the septal wall, Prieto manipulated the catheter to its peak and, after several *we're-going-to-start-again*s, pressed the needle carefully, carefully through. She squeezed some agitated saline through the needle—it looked like little bubbles on the echo—to verify that it was in fact in the left atrium. It was. Now, by pressing a syringe filled with water, she could inflate the balloon, then let the water rush back into her syringe, leaving a hole in the atrium. Adel registered a jet of motion between the septums: flow. She ballooned again, enlarging the hole. The oxygen saturations immediately shot up to 50 and then 60. She ballooned three times. Blood could now leave the lungs. Saturations rose to 80 percent. Adel recorded good flow through the newly made hole. Prieto retracted the catheter. The repair had gone just as hoped. The baby, Kenneth Mangan, had suffered serious lung damage, but his oxygenated blood now had a pathway to his body and could feed his tissues. That would in turn allow the intensivists to stabilize him in preparation for, they hoped, his Norwood.

• • •

Later in the day I stopped Prieto as she passed me in the same slouched, expressionless gait she'd used when she headed to the O.R. to ask Mee's advice, preparing for the procedure in her head. I asked for her thoughts. Her only comment was, "I was lucky this kid had good anatomy."

She recalled the postmortem on the last baby: among the findings had been copious blood in the mediastinum, suggesting they'd punched through the back of the atrium. Just then Mee walked up in his white O.R. trousers and white short-sleeved shirt. He mentioned a few more fatal cases, including one here, in which the cardiologist had pulled on the balloon and ripped the septum out of the roof of the heart, and others elsewhere in which the needle had punctured the back of the atrium—"That's what I was afraid of doing," Prieto said—or the sats had been so low the baby was overventilated, which destroyed the lungs instead of elevating the blood oxygen levels, or the blood had soured as the organ tissues ran out of oxygen.

Prieto nodded and then slouched off down the corridor. Mee scratched his belly.

I asked him how *his* case had ended up going. He described what he'd done and admitted, in an effort at modesty, "I'm quite pleased," but he couldn't suppress a shit-eating grin after he said it. (It had been perfect.)

Mee and Prieto could both go home tonight knowing they'd put in a good day's work: two children's lives saved, and the lives of their families transformed, a red-letter day.

Kenneth Mangan, however lucky he might be, remained in a precarious state, his little lungs having sustained a substantial blow. Not yet twenty-four hours old, he'd already been through a lifesaving procedure, and the hard part was still a week away—provided, that is, that the intensivists could keep him stable and free of infection or fever until then. Patricia Mangan would ask Mee if what her baby had been through would make the Norwood operation more difficult. Mee would respond that he couldn't answer that, because he'd never known a baby with Kenneth's defect to survive.

. . .

Hypoplastic left heart syndrome—the defect that Kenneth Mangan was born with and that another baby, a girl named Sommor Beeman, would be born with two days later, and Matthew Anderson a week after that, all of them at the clinic—was the last major hurdle in the history of congenital heart surgery to be overcome by daring, innovative surgeons. While today the defect is no riskier than several other complex heart defects, and more and more physicians, especially given the increasing success of transplantation, find themselves uncomfortable providing no treatment to babies born with it, the very notion of being able to *fix* hypoplastic left heart syndrome, a heart missing its main pump, once seemed inconceivable. About a thousand babies are born each year with the defect (according to one conservative estimate), making it among the most common of heart defects, as well as one of the most severe. For virtually all of human history, HLHS babies simply died shortly after birth, when their ducts closed off. The more talented echocardiologists began to diagnose it in utero in the early 1980s and mastered its diagnosis by 1986; shortly thereafter it could be routinely picked up by an obstetrician's ultrasound, given a skilled practitioner with good equipment (ability and equipment would continue to be inconsistent for years, however, and even today hypoplasts often go undiagnosed until after they're born). Once the diagnosis had been made, the doctor could offer termination of the pregnancy, or if it was too late for that, or not an option for the family, the parents could be advised that the baby would be kept free of pain as everyone waited for his or her death.

In the late 1970s, when some form of repair or palliation was available for virtually every other type of lesion, babies born with this fairly common but fatal defect "were being pushed away in the corner to die," according to Bill Norwood, "and they were otherwise perfectly normal."

Norwood was a young pediatric heart surgeon at Boston Children's when Roger Mee arrived in 1977 for his six-month rotation under Castañeda. Mee, a few years his junior, recalled that even then, six years before the publication of his landmark paper in the *New England Journal of Medicine*, Norwood was already trying to devise a repair for this seemingly irreparable defect.

"The principles of managing hypoplastic left heart syndrome, the central principles are—you can dream 'em up pretty easily," Norwood said from his home in Delaware, where he continues to practice. "You need to have the aorta somehow permanently connected to the right ventricle. You need to have a wide-open interatrial communication because otherwise blood comes back from the lungs and has nowhere to go. And if you do that, somehow you have to limit pulmonary blood flow. So you go to your repertoire of things to do to accomplish those structural changes, and combine them."

Here Norwood was following the logical method that had brought congenital heart surgery so far in its ability to treat very complex lesions. At first, he said, surgeons concentrated on structural repairs of the heart, on the actual plumbing, on making a defective heart *look* like a normal heart, but that turned out to be secondary. "It wasn't until the early to midseventies that the really complex lesions seemed fixable," he said. "We began to realize collectively—there weren't many of us—that the holy grail was really physiology, not structure." This conceptual leap from structural fixes to the physiological ramifications of structural changes "was a major, major breakthrough."

Both Mee and Castañeda remembered Norwood as being a "frustrated surgeon" at the time, because all the cardiologists wanted to send their patients to the world-famous chief surgeon—his boss, Castañeda. Norwood had a lot of time on his hands, so he started thinking about HLHS. By 1979 he had enough of an idea worked out on paper that Castañeda agreed to let him try it. Unlike the go-ahead for the first neonatal switch several years later, this decision wasn't a morally wrenching one. With that later case there would be a low-mortality alternative to the new procedure, whereas here the only alternative was certain death. The baby was going to die anyway, so Castañeda authorized Norwood to give it a roll. By Norwood's account, the results were good from the beginning—a 30 to 35 percent mortality he said—but others recalled the mortality as being awful and discouraging as the team tried to figure out what worked and what didn't. As with every innovation, there was a learning curve.

Asked about this, Castañeda chuckled and said, "Bill's memory is failing. He started out with different ideas. The present Norwood operation is not what he started out with." The first Stage One palliations, which Norwood would describe in 1980 in the *American Journal of Cardiology,* worked on two out of three patients. The procedure involved the insertion of a tube to connect the right ventricle to the descending aorta, as opposed to the pulmonary artery/ascending aorta connection that is used today. Castañeda explained that because this conduit required that the only working ventricle be sliced into, the results were not good: "There were a number of operations and failures. It was a busy, tough going."

The development of the Stage One palliation was an evolutionary process, Castañeda said. He was never against attempting it, but he wanted it to be Norwood's project. "I did think it was interesting, but I left it to Bill," he noted. "I stayed out of his way."

Beyond its technical aspects, the repair's success relied on a *system:* "It wasn't only a surgical technique," Castañeda concluded. "It had to do with preoperative management, intraoperative management, and postoperative management, so it's a whole teamwork. Anesthesia is extremely important."

In talking to me, Norwood minimized Castañeda's direct involvement: "Aldo was happy to see me doing something; I don't think he was particularly interested in it, nor did he particularly understand it."

Mee's recollection of Stage One's genesis was more colorful. Norwood worked it out for Mee. They discussed it, what would be necessary for the palliation to be successful "over a beer at Children's Longwood Inn." Mee was then asked to describe it in conference, whereupon Castañeda said he'd allow the experiment "over my dead body." According to Mee, Norwood kept bugging Castañeda until he relented and agreed to let his young surgeon try it.

I asked clinic anesthesiologist Paula Bokesch for her recollections. She'd started at Children's in 1981 as a fellow and was part of the team that performed the first successful complete repair in May 1982—the first baby born with HLHS to survive the final surgery, and the future subject of Norwood's groundbreaking paper. ("It was a big day," she said. "I remember Aldo said, 'History is being made today in room ten.' Bill stayed with the patient at the bedside all night.") Bokesch would continue to work with Castañeda and Norwood for the next fifteen years, in Boston and again when they opened a center in Switzerland, after which she would come to Cleveland to work for Mee. She knew well and admired all three surgeons.

"In the 1980s, as again in Switzerland," Bokesch wrote to me in an e-mail, "Aldo had all the cases referred to him except the hypoplasts. Bill was doing one case a week and . . . needed something to keep him occupied. . . . Knowing both of them well, I'd say it was probably Bill's idea—or at least Bill definitely pursued it against all odds. Also, Bill is very smart and analytical. It is typical of him to sit for hours and draw ideas. He also designed a golf club. He's a physics nut."

I wondered if Mee wasn't trying to horn in on Norwood's legend, maybe pull the water out from under his feet a little, since Norwood himself didn't recall any boozy brainstorming session. Still, I loved the notion that the repair had originated in Norwood's imagination and then been drawn out on a cocktail napkin, the two thirty-something surgeons discussing it over beers—it seemed like a throwback to the 1950s era of heart surgery. So I asked Mee again. When I told him that neither Castañeda nor Norwood remembered it his way, he was peeved. He described exactly where he'd been sitting in the conference room that day (left-side seat, third row up) and said he hadn't even wanted to mention it until someone else had asked what was to be done about the hypertensive left ventricle. Mee was called on to respond and the room subsequently fell silent. The idea was shot down ruthlessly. "I felt quite a fool," Mee recalled. Given that he left Children's Hospital in June 1977 and Norwood did his first experiment the following November, the timing makes his story plausible.

Richard Van Praagh, a pathologist and cardiologist at Children's until his recent retirement, who was there at the time, doesn't recall this particular conference, but he notes that Roger was more likely the spokesman; Norwood did not speak comfortably in front of his colleagues (Norwood once remarked, Van Praagh said, that he ought to have a personality transplant). Van Praagh recalls, "People were very much afraid. But that was the reason it had to be done.

Aldo was not at all enthusiastic about it." Furthermore, he notes, "The mortality was considerable." But more interesting, it was Van Praagh who had devised a precursor procedure six years earlier for a different lesion, interrupted aortic arch. Van Praagh's procedure accomplished many of the same things the Norwood procedure would: an interatrial communication, a pulmonary-aortic conduit, sufficient pulmonary arterial stenosis. It wasn't intended to be used for hypoplasts, but since before Norwood arrived, it had been part of the "intellectual ambience" of Boston Children's. "What Bill Norwood did was much more elegant than what we did," Van Praagh added.

No one, of course, disputes Norwood's right to the name of the operation (though some centers, including Boston, refer to it as Stage One), nor does anyone deny full credit for it to this surgeon who invented a surgical repair that now saves countless babies throughout the world, through his extraordinary effort, ingenuity, daring, skill, and willingness to accept moral and legal responsibility for surgical mortalities and, worse, the possibility of late mortalities, children who would struggle along until they were two or three and only *then* die. The very first patient died seven hours postop. Norwood attempted the procedure again two months later, and again a month after that, and again another month after *that;* this fourth patient was extubated after three days and left the hospital after three weeks.

Keeping the hypoplastic heart alive was uncharted territory, and the long-term results relative to quality and length of life were—*are*—still unknown in any statistically significant way. In the early days many nurses and intensivists protested the intervention, convinced that the high mortality made these experiments inhumane. Today mortality remains higher in Europe than in the United States. One center in the United Kingdom recently published a study of eighty-seven cases of HLHS diagnosed in utero: of these, thirty-eight were terminated before birth; eleven were born but received no treatment and died soon after; thirty-six had Norwoods, with twelve surviving (giving a 66 percent mortality for the procedure); and two died of other causes. Smaller centers in the

United States still have high mortality, and many don't offer the surgery at all because of that. Some centers still advocate termination of pregnancy or compassionate care rather than referring a mother to a larger center and committing both a child and family to three open-heart surgeries within the first four to six years of life—that is, in the best-case scenario. Nevertheless, I've met parents who held their children—babies born with HLHS—in their arms with gratitude and couldn't imagine a world without them in it; to them, *not* having at least the option of a Norwood would be unthinkable. I have heard a fiery seven-year-old who'd been through all three surgeries demand to know what kind of book I was writing, anyway; when I asked her some questions, she bashfully hid behind the leg of her mom—a beautiful child, and by every visible measure in vigorous good health. And Bill Norwood, standing on the shoulders of Gross and Lillehei, of Gibbon and Kirklin, of Castañeda with his advances in neonatal surgery, was the man responsible.

But even Norwood himself, at a conference in the spring of 2001, noted that the procedure was still "a technically demanding operation on a quite tiny baby," and conceded that through the 1980s and most of the 1990s it had been considered "illegitimate." When Mee began performing the surgery in Australia, he encountered severe resistance from his ICU team, some of whom he believed actually hindered the care of babies who'd had their Norwoods. Some doctors are ambivalent. Marco Cavaglia is now one of them. An anesthesiologist and the son of an Italian heart surgeon, he'd worked with Castañeda and Norwood in Switzerland and with other surgeons throughout Europe; he was at the Cleveland Clinic during my time there, assisting his old colleague Paula Bokesch with a study on the effects of cardiopulmonary bypass on the brain. Cavaglia said he remained skeptical that the Norwood should be offered at all. The mortality was 30 percent, he said, and the quality of life for those who survived wasn't good. When I said I'd been under the impression that mortality was lower (Mee's was 16 percent, and Michigan's was even lower than that, for instance), he smiled cynically. On the subject of mortality figures he cautioned me, "Don't believe anything you read," explaining that such data were too easily manipulated, and that in any event each case was so different as to be impossible to compare with others in any meaningful way. "Not a good result, not a good result at all," Marco said. "Have you seen a fourteen-year-old Fontan?" No, I said.

Because no national data are kept, precise figures on hypoplasts are unknown—but if they constitute 7.5 percent of the approximately 32,000 babies born here each year with heart defects, then we can estimate a yearly total of 2,540 hypoplastic newborns. Other physicians and researchers have put the incidence at between 2 and 2.6 hypoplasts for every 10,000 live births, a statistic that reduces the total to 800 to 1,000 per year. In a given year we don't know how many of those, say, 1,000 will have Norwoods, how many of *those* will go on to survive their second and third operations, or how many children and young adults are alive today without a functioning left ventricle. Many palliated hypoplasts still die; most of these deaths come shortly after the Norwood, but the worst of them wait until after the second operation but before the third, when the child is a toddler, calling for Mommy, asking for juice, becoming a little person. "That's really tough," Mee said, "really tough." As more hospitals offer the surgery, as more doctors begin to feel it's unethical *not* to recommend it, as results continue to improve, more and more parents are choosing surgical repair.

Results still vary widely. A recent study gathering data from more than sixty university hospitals that offered the surgery from 1990 to 1999 found that of the 2,264 patients admitted with HLHS during that period, 1,203 had Norwoods, with an overall mortality of 42 percent (though the five centers that averaged more than fifty Norwoods apiece per year all recorded lower mortality than that). Ed Bove and his partner Rick Ohye, at the University of Michigan, reported a 14 percent mortality for standard-risk patients and a 58 percent mortality for high-risk ones (those weighing below 2.5 kilograms or having associated lesions). Ohye noted, however, that these numbers were about ten years old and thus dated; overall mortality had continued to decline, he said, and now stood at about 15 percent for all patients undergoing the Norwood procedure.

. . .

Gwendolyn Ferchen lies on the examining table, her soft, broad belly slicked with lubricant and exposed for the echo wand. She is on the eighth floor of the Cleveland Clinic's A building, in the echo room, where she has just been examined by her new OB-GYN, Stephen Emery. Emery, who specializes in fetal treatment, now introduces her to Adel Younoszai, who has come to echo her babies' hearts. Seated just off her left shoulder is her husband, Stephen, who works for General Motors in Buffalo, the city they've traveled from to get here today. He will watch the echo on a big screen above Adel. Gwen is carrying twins. Her doctor in Buffalo, Joseph Orie, recently diagnosed hypoplastic left heart syndrome in one of them and recommended four centers—Boston, Philadelphia, Cleveland, and Michigan—for further treatment. Confirmation of that diagnosis and meetings with their new team are the reasons for the Ferchens' visit.

The nurse guides the probe over Gwen's abdomen. Gwen wants to deliver naturally, and she asks if that'll be possible. Adel can't answer her question; he says that when the time comes, her doctor will make the decision, adding, "There'll be a lot of people in the delivery room. There's going to be an anesthesiologist, and there are going to be two teams, one for each baby." He pauses as the nurse finds the first heart. "Perfect picture," Adel says.

Four chambers are distinct—two atria atop two ventricles, beating in perfect synchronicity. Adel scrutinizes the valves and says, "You can actually see the hinge." Then he says, "This baby's heart is beautiful."

"Thank you," Gwen says. "His name is Ethan."

"What's Baby A's name?"

"Evan," she says.

The nurse guides the probe over Baby A, Evan, who is oriented head down, first into the starting gate. "He's got his hand in the way," the nurse says.

Adel says, "You can try coming down here and angling this way." When the nurse continues to have difficulties, he asks, "Mind if I have a shot at it?" The nurse steps aside, and Adel takes a seat on the stool.

Gwen says, "Harder with a porker."

Adel chuckles and smiles. "In this case size does make things a little more difficult," he agrees.

Gwen carries a few extra pounds that probably aren't related to her pregnancy, but she's tall and so wears them well. She's a pretty woman, twenty-seven years old, and has a sweet, soft voice that now trembles a little with what seems to be a mixture of excitement and fear.

It takes Adel several minutes to get a good view of the heart. "There's a four-chamber view," he says softly for all those watching the two screens—besides himself, Dr. Emery, the Ferchens, and the nurse. "Large right atrium, large right ventricle, sliverlike left atrium, sliverlike left ventricle. There's the ventricular septum." He keeps his eyes on the staticky black-and-white images of the echo, which displays a two-dimensional layer of the heart at any given instant (three-dimensional echo is still being developed). "Pretty decent-sized ascending aorta," he remarks, measuring it with a push of a couple buttons. "The only thing we haven't seen is the arch, but both ascending and descending look good." Adel glances from the screen back to Gwen and asks, "Are you uncomfortable?"

"Yeah, but that's all right."

"I don't want you to be uncomfortable."

"It feels like my belly button is about to pop," she says. The combined weight of the babies, the nurse has estimated today, is twelve pounds and growing—Gwen's got six weeks to carry to term.

Evan remains in a difficult position, so Dr. Emery holds a vibrator against Gwen's abdomen to encourage the baby to move. Adel has by now checked and measured most of the structures of the heart—the atria, the ventricles, the ascending aorta and pulmonary artery, the tricuspid and pulmonary valves—but he has yet to find the arch, and he can't determine whether the mitral and aortic valves are open or even extant. Gwen senses his disappointment and asks pointedly, "Is everything all right?"

"There's no problem," Adel says. "So far, what I see confirms what Dr. Orie found." He concludes the echo and wipes her abdomen with a towel that he then leaves draped across her. "How much did Dr. Orie tell you? Did he talk about—"

"He told me about the options. Surgery is definitely . . ."

"OK, good," Adel says. "If we can make certain measurements, we can give you an idea of whether there will be additional risks. Right now I don't see any, but we won't know for sure until the baby's born. Dr. Mee will evaluate the situation then and decide when will be the best time to operate."

"How many has Dr. Mee done?" Stephen asks, still seated.

"I don't know his exact numbers. That's something he can tell you. *I* can tell you his success rate is as good as anyone's in the country. And if I had a child, I would not hesitate to send him to Dr. Mee." Adel pauses and then reiterates that these pictures are still preliminary. The important thing today, he says, is that they've confirmed the diagnosis.

"Most people feel guilt," he tells the couple. "There's nothing you did that caused this. Some people think, I shouldn't have taken this medication, we shouldn't have had sex, I shouldn't have eaten that piece of fish. There's nothing you could have done to prevent this." He pauses again and asks, "Do you want to have more children?"

Gwen nods and says, "Yes."

"You're at slightly greater risk for having another baby with heart disease. *Slightly*. If there's no congenital heart disease in the family"—both Gwen and Stephen shake their heads no—"normally about one in every hundred babies will have congenital heart disease. For you the risk goes up to three in a hundred.

"I do need to warn you that when these babies are born, they look perfectly normal. But a small number, after they're born, their disease is so severe that there's nothing we can do for them."

Following more discussions with the obstetrician, Gwen and Stephen are directed to the M building, fourth floor, a ten-minute walk through the behemoth clinic complex. They wait for over an hour beyond the time of their scheduled meeting with Dr. Mee, who's gotten behind with other consults and a visit from three valve salesmen. Gwen and Stephen have been escorted to a small examining room (padded examining table, carpet, sink, devices affixed to the far wall to measure blood pressure and look in ears and eyes, two adult chairs and one toddler chair—generic but not cold).

Mee finally enters carrying a folder, looking a bit sheepish about his lateness. He apologizes. It's a Monday afternoon, so he's wearing his dark jacket and tie, black trousers. He glances around the room, sees that the only adult chairs are already occupied, and so leans across the examining table, opening a manila folder.

"Are you holding up all right?" he asks. The couple both exhale and nod. "Yes? Blood pressure all right? Good."

Gwen and Stephen are big folks, solid American stock, and they sit side by side, their shoulders close together. Stephen wears glasses, is balding, and has hair that's clipped short. He's quiet, letting Gwen do most of the responding. The first issue Mee brings up is the duration of her pregnancy. He wants the babies to stay in there as long as possible; low birth weight, he explains, is a risk factor. Conventional wisdom was once in favor of inducing early—at thirty-six weeks—but he says that's changed. The thought of these babies getting bigger inside her, the thought of six more weeks, makes Gwen shake her head, but she understands.

"Do you know the sex?" he asks.

"They're both boys," she says.

"Have you named them?"

"Yes, Evan is the baby with hypoplast."

"Evan and . . . ?"

"Ethan."

Mee's brow furrows. "Is that a Welsh name?"

"Yes, it is," Gwen says, pleased that Mee knows this. "I loved it as soon as I heard it."

"How much have they explained to you?" Mee asks. "Have they discussed all your options? Have they told you about the option of doing nothing?"

"Yeah," Gwen says quickly, nodding. "We do want the surgery."

"You've thought about that, and you're both in agreement?"

They say yes together, and Mee says, "That's important."

Next he moves on to the risks. He notes the good size of the aorta, reading from Adel's echo report. "When it's less than two-point-five millimeters, the risk is quite high—about thirty percent. Greater than that, the risk goes down to three."

Now Stephen speaks: "Is that your statistic, or everyone's?"

"That's ours," Mee assures him.

Stephen's question is an important one—it's not unheard of for a doctor to cite general statistics rather than personal results.

"The definitive information will come after he's born and he's echoed," Mee says. "In the old days—well, not so old—they were born and they looked good, they looked perfectly normal. And then within twenty-four hours hell broke loose and they quickly died." Gwen and Stephen need to be prepared, he reiterates, for the baby to look healthy.

"In your opinion," Gwen asks, "can they make it through a vaginal birth?"

"Oh, yeah, they usually do OK." He notes a couple of potential problems, such as the intact atrial septum. "That's uncommon, real uncommon, but it happens." He mentions a few other problems, and then, tired of bending over the examining table, he sits in the toddler chair, whose seat is about six inches off the floor.

"I'm so glad I'm here," Gwen says.

As if sensing that comfort isn't appropriate here, Mee launches into a series of warnings about the risks of surgery, telling the couple how the first operation is by far the most serious; how the heart will still have to work doubly hard until the second operation, in four to six months; how the lungs may become badly hypertensive, making the second stage impossible; how, even if the baby makes it through the second stage, the right ventricle may deteriorate, precluding any possibility of the third and final stage. "Transplantation is part of the management plan," he adds. "You should square that away in your minds."

Gwen says, "We both want to do as much as we can to make him healthy."

"Transplantation is a palliative operation—it doesn't make your child normal," he warns, reciting the litany of daily antirejection drugs, possible rejection, coronary disease. But for now there's nothing more he can tell them. He'll evaluate Evan when he's born.

"When will we see you?" she asks.

"You'll be seeing a lot of others before you see me, but not too long afterward, I'll come by to pontificate."

Gwen takes a breath and says, "It all seems so crazy."

Mee nods.

Stephen asks, "How many of these operations have you done?"

"I haven't done as many as a couple of people in the country," Mee says. "Ed Bove in Michigan has done quite a lot, and the people at CHOP in Philadelphia. I've done about eighty."

"Eighty?" Gwen says, apparently surprised by the number.

"Yeah. Our results are about the same as CHOP's and Ann Arbor's."

Stephen says, "I'm worried. Is he going to feel this?"

Concern over pain is common among parents, so Mee has an answer ready: "We don't think pain is good for them—so no, not much pain." The team has sophisticated ways of managing that, he says.

"How long will he be in the ICU?"

"Anywhere from two to three days to over a month," Mee says, and then pauses before speaking again. "One thing you need to know is that the physiology is the same before the operation as it is after. There is still one pumping chamber to pump to the lungs and the body. That's quite a lot of pumping. Things can go suddenly bad at home—sudden death." If that happens, he doesn't want parents to feel guilty about having taken their baby home too soon. "After the Glenn," he goes on, "it's a much more efficient system." He stresses that while the three operations—the Norwood, the Glenn, and the Fontan, done in that order—work to make the heart as efficient as it can be, the heart of an HLHS baby will never be normal. This raises the issue of transplantation again, and again he emphasizes, "It's another form of palliation. It can go well, or it can be a lot of trouble, with chronic rejection, infection. You shouldn't have any illusions that transplant will solve everything."

As he always does, he underscores the physical limitations Evan will face later in life. Highly competitive sports will be out of the question, so if they have any hopes in that area, they'll need to rethink them. Stephen says, "I don't care what he does, as long as he's around."

"I did nothing but sports, and I wish to hell I'd learned to play the piano," Mee says.

Stephen asks, "How far away are we from using mechanical hearts?"

Mee tells him that so far they've been used only for brief periods, as a bridge to transplant. They're not good enough yet for anything more than that, but that may change soon. Xenografts (nonhuman hearts), too, are being heavily researched. "Both forefronts are moving ahead," he notes.

Mee gives the impression of being in no hurry. He doesn't look completely comfortable in the toddler chair, but he seems at ease. He asks when they got the first diagnosis.

Gwen tells him, then says, "The initial shock is . . . when we first found out . . ."

"It's a lot to cope with," Mee says.

"But I wouldn't trade it for anything."

"It really is a blessing, fetal echo," Mee says.

They converse easily—about Gwen's earlier miscarriage, how hard it was for her to get pregnant. Mee tells them a bit about the cardiology team and explains that the cardiologists all live close to the hospital, adding, "We talk a lot between ourselves." Finally, he says she'll need to keep herself together, too, and shouldn't feel bad, when the time comes, about leaving the baby in the ICU so she can get some rest.

"This is a bad deal," Gwen says. "But I really believe we're in the best place possible."

Mee bids them farewell and encourages them to call him with any questions. Gwen and Stephen stand to leave. The day is almost done, and Gwen's only surprise is that she won't be giving birth soon—she thought they'd want to get things under way immediately. Now she and Stephen have six more weeks to reflect on the unknown, on the birth of their twins, one with a dangerous heart defect.

Cindy and Doug Beeman, like Gwen and Stephen Ferchen, have decided to opt for the surgery rather than compassionate care.

(More than six weeks after her first meeting with Mee, Gwen would be lying in a spacious birthing room, hooked up to an IV filled with a drug that would induce birth. Adel would stop in to say hello and wish the foggy but coherent mom-to-be well, and on leaving he would say softly, "It's the start of a long journey." Those words would sound so ominous to me that I'd ask him what his thoughts were on compassionate care—did he feel it was still a viable option for Evan Ferchen? "I believe it's an option all the way up until he goes in for surgery," he would respond without hesitation.)* The Beemans' daughter, Sommor, was born two days after Kenneth Mangan, under decidedly less dramatic circumstances. Sommor was to be the perfect hypoplast baby.

Cindy's obstetrician hadn't noticed anything unusual during her first ultrasounds, but in week thirty-six, when he still couldn't see all of the baby's heart, he sent her to a specialist at Children's Hospital in Akron, not far south of Cleveland. After an interminable and deathly silent echocardiogram, the cardiologist confirmed her worst fears: she had a very sick baby who might not live. Cindy admits that her initial thoughts were not for the baby who was due in a few weeks, but for herself. She began to cry, feeling as though she'd plunged to some rock bottom. "How could this happen to me?" she wondered. "How could this ruin my family?" A previous pregnancy had ended at eight weeks, in a Christmastime miscarriage. And now this—"another devastation," she says.

A twenty-nine-year-old mother of two—a four-year-old and this new, sick baby girl—Cindy has soft features and straight dark-brown hair and is a teacher with a master's in instructional technology. Her story is not an atypical one at the clinic: upon diagnosing

---

*When I contacted Dr. Younoszai the following year to check some facts and verify this statement, he responded with an e-mail that was even more ominous, but in an unexpected way: "It's interesting to reflect on that quote. Today, as opposed to a year ago, I have, perhaps, become somewhat more optimistic about the outlook for patients with hypoplastic left heart syndrome. I have now begun to see what they look like as they grow into children. I think there is a general trend that way in the pediatric cardiology community with the surgical results so improved over the last ten to fifteen years.

"Ask me again in fifteen to twenty years, when these children (the data predict) are in end-stage heart failure, and I may feel otherwise."

HLHS in her unborn child, the Akron specialist referred her to Dr. Mee. "He didn't say, 'Dr. Mee is God,' he didn't say, 'He's the best in the world,' he didn't say, 'You are so fortunate to live so close to him,' " Cindy recalls. "I figured the guy's got such a short name, he can't be anything. I pictured this tall, tall guy—stalky, mean.

"I was scared to death of Dr. Mee. And the day of the appointment, in walks this little short guy, a little pudgy guy, and I instantly fell in love with him and relaxed completely, and my husband put his hand on my knee, and that was his signal that everything was going to be all right."

In their consultation Mee spoke to the Beemans as he would to the Ferchens, noting risks and making sure someone had talked with them about the option of compassionate care, of doing nothing.

When Cindy's labor began, her husband, Doug, did the driving. Doug, also twenty-nine, is an accountant and a creature of routine. "He drives sixty-three in a sixty-five zone," Cindy says. "I said, 'You can kick it up to seventy because the contractions are now four minutes apart.' " But she tried to remain relatively calm so as not to upset him. *He* was most worried about losing Cindy, while *she* was certain she would lose the baby. She hoped at most "to give her a kiss and say good-bye."

But when Sommor had emerged after just ten minutes of pushing, and Cindy was holding her in her arms, the baby reached up and touched her face. The delivering doctor said, "Cindy, you don't get hypoplasts that look this good." She and Doug cried, overcome with joy that their daughter was here and would be all right, that their family was, for the moment, whole. Everyone was so positive, Cindy says. No one even mentioned compassionate care after the baby was born.

The three mothers, Patricia Mangan, Cindy Beeman, and Tera Anderson—all of whose HLHS babies are currently in the unit—talked. Tera's son had been born a week after Cindy's daughter. Before the surgery, the weekend after the birth of her baby, Tera asked Cindy, *When your daughter was born, did they come to you and talk to you about compassionate care?*

Cindy paused, then lied: yes, she said.

("I've had to hide so much good news," she'd tell me later. Sommor had had trouble coming off the ventilator, but that'd been the worst of her problems: she was out of the ICU in three days and recovering as well as any hypoplast the nurses had ever seen. Cindy couldn't help but feel guilty—why was she so lucky, why was her *daughter* so lucky, while Tera and Patricia and their families continued to suffer?)

Tera's course had been virtually identical to Cindy's. She had gone to her doctor for a routine ultrasound at thirty-four weeks to check the position of the baby. Her doctor, unable to view all four chambers of the heart, had sent her to the same specialist Cindy saw. This doctor diagnosed hypoplastic left heart syndrome and named three places that could do the necessary surgery: Rainbow Babies' and Children's in Cleveland, the Cleveland Clinic, and the University of Michigan. Tera is a tall, sturdy woman with light-brown hair, then the mother, like Cindy, of a four-year-old—a boy named Phillip. Phillip's father, Anthony, age twenty-three, worked as a mechanic for Davey Tree Service, repairing bucket trucks and motors. Trim and with his dark hair and beard cut close, Anthony was a quiet, thoughtful man. He and Tera wanted to do all they could for their second son. Tera held Matthew for a few minutes after he was born, and then the doctors placed him in a warmer bed and rolled him to the PICU. "It was good to be able to hold him," Tera said.

Within twenty-four hours, however, the baby began to have problems. His pulmonary circuit was not efficient, and he had to be put on mechanical ventilation on his second day of life. Then came more bad news. The size of a healthy newborn's ascending aorta— the section of the aorta that rises out of the left ventricle—depends on the baby's weight, but it typically ranges from six to ten millimeters. The aortas of most hypoplasts are between two and four millimeters. An echo measured Matthew's ascending aorta at less than two millimeters, which Mee has cited as a major risk in the Norwood procedure. All things taken into account, Matthew was scheduled for the earliest slot on the following week's surgical schedule. Mee quoted Tera and Anthony a 25 percent risk of death, higher than normal because of the size of their baby's aorta.

• • •

When Mee entered O.R. 51 to adjust his fiber-optic headlight, Fackelmann, who was assisting Charlie in opening, relayed the news that Matthew Anderson's aorta was indeed unusually small. Mee left the O.R. to scrub just outside the door, banging his knee into the steel sink to start the water. I'd planned to observe today and was leaning against the desk in the large vestibule. Chafing at his burly forearms and elbows and wrists and fingers with the soapy yellow brush, Mee was clearly concerned. Through his head-gear, in the mirror above the sink, he caught my eye, shook his head, and said, "This could be a real bumfuck."

Julie Tome, the anesthesiologist for this case, was likewise concerned, aware that unless the baby was managed perfectly, there could be trouble: his aorta was so small the heart could arrest at any moment during the manipulations of sedation and intubation. It's rare these days to lose a patient in the O.R., but given the anatomy today, it was possible. Mac, Mee's senior fellow, was away visiting family in Japan, so Mee had made sure Fackelmann was available to assist. He asked Charlie, an M.D. and a surgeon, to switch places with Fackelmann. Charlie asked to be given a chance and Mee said no: he wasn't ready. I'd seen how Charlie's hands shook in tense situations, knew he could be visibly uncertain in his movements. Mee had no tolerance for such things. (In another case this week Mee would say across the patient's open chest, "You're nervous, Charlie. I can see what you're thinking. You're thinking, Am I good enough to do this? You can't think about yourself. If you're thinking about yourself, you're *not* good enough. Everything has to be directed at the patient.") Mike took Charlie's place, directly across from Mee.

Mee was shocked when he saw the heart and the aorta for the first time. "I've never seen one that small," he said. It was half the size he'd expected—a millimeter, he guessed, thinner than a cocktail straw—and with little more reflection than that, he began cannulating the ductus, the vessel through which freshly oxygenated blood would be pumped to the body.

• • •

The classic Norwood operation comprises three main steps: transecting the main pulmonary artery and sewing it into the aorta; sewing in a shunt, a 3- or 3.5-millimeter plastic tube, between the aorta and the pulmonary artery to deliver blood to the lungs; and opening a large hole in the atrial septum. As Norwood himself noted, it's a tricky operation, on a very small heart. Because the aorta is smaller than normal in HLHS babies, and because the surgeon must fabricate what is in effect a new aorta arising out of the pulmonary artery, extra tissue is typically used to form this new vessel. Traditionally most centers have opted for something called a pulmonary homograft, a main pulmonary artery harvested from a donor and kept frozen until it can be used. From this the surgeon cuts a large piece of tissue, a patch to connect the pulmonary artery with the aorta and its arch. It's a complex shape, a rectangle wider at one end than the other, which must bend in two directions at once; the entire aorta is shaped like a walking cane with a rounded handle, the arch. The patch must be in a shape that can be sewn into the underside of that arch.

This method never satisfied Mee, however. He liked to avoid homografts whenever possible because he'd found they could calcify and harden over time, resulting in complications in the ensuing procedures and afterward; he preferred instead to use only native tissue. The patch also requires a lot of sewing—sewing, moreover, while the heart-lung machine is turned off, a state called circulatory arrest, a race against the clock to prevent brain damage. He hoped to cut this circ-arrest time, the period during which the baby's brain received no oxygen, by reducing the number of stitches thrown. He'd figured out a way, by manipulating the descending aorta—the long shaft of the cane, as it were—dissecting it out, and tugging it upward, and then making simple incisions in the aorta and pulmonary trunk, to build a new aorta without homograft tissue: an all-native-tissue Norwood, or what he called a modified Norwood, a description of which had recently been published in the *Journal of Thoracic and Cardiovascular Surgery*, accompanied by Mee's own drawings.

It was an ingenious bit of cutting and sewing, but only a few surgeons had adopted his method, including Mee's former fellows

Bill Brawn, in Birmingham, England, and Nancy Poirier in Montreal. Many others were aware of the modification but had evidently chosen not to switch over from the old method, which after all worked just fine. Joe Forbess in Boston asked me about Mee's procedure, openly curious and intrigued by the idea of an all-autologous-tissue repair. After the clinic's new surgeon, Brian Duncan, observed Mee's technique for the first time, I asked him what he'd thought of it. He said, "Sweet, man"—he especially liked the reduced amount of sewing it required. (Duncan would nevertheless continue to use the method he'd learned in Boston, employing a homograft patch.) I asked the Man himself, Dr. Norwood, for his thoughts on Mee's modification of his operation. In a tone that I can only describe as surgeonlike, Norwood said, "I have no idea how Roger does his, and I don't care because I know what I'm doing is right."

All of these surgeons agree on one thing, though: the procedure's requirement for perfection. The baby's heart and pulmonary circulation are so precarious immediately following the procedure that there's little room for error. A redone patch, an unnecessarily long circ arrest or pump run, inexact stitching—a single-ventricle heart doesn't tolerate such business.

"It really takes a whole team to do," says anesthesiologist Paula Bokesch. "And any mistake on any one person's part is disastrous. That is the one operation where there are no unforced errors. When you do a hypoplast, it's the finals at Wimbledon."

An error was precisely what happened early on in Matthew Anderson's operation. The O.R. was quiet and orderly, as always—the only noise the hum of the refrigeration unit of the pump, the beeping of the monitors. It was almost always the same; Mee's routine for cannulating—getting the heart-lung machine's tubes into the proper vessels—was unvarying; an observer could set a watch by it in most cases. So at any variation in the routine, or any uncharacteristic remark from Mee, everyone's ears would naturally prick up.

"Shit, I got it," he said—"it" meaning the aorta, the vessel that perfuses the whole body and, most critically here, the heart itself. It was blown—a potential catastrophe for baby Matthew, whose five-day-old chest was spread open beneath the bright lights of the O.R. With a perceptible tremor and unusual urgency in his voice, Mee said, "We're in trouble, guys. We've got to get on pump quick."

Fackelmann glanced at the monitor, then down at the heart, then at the monitor again, but was otherwise still, sucker and pick-ups in either hand. Mee began to cannulate quickly.

Then something more ominous happened: nothing. The aorta bled a little, but that was all. The heart remained bright red and continued to beat without interruption. This was perhaps even scarier than getting into the aorta before they were on pump; the heart should have reacted to this, but it didn't. Within a few minutes Matthew was safely cannulated, and there were vague mutterings between Mee and Fackelmann about why nothing had happened. The heart evidently wasn't relying at all on the aorta but must have been perfused by other means, maybe collateral arteries, which might explain why the aorta was so tiny as to be almost useless. And if this was the case, what did it mean for little Matthew? The case proceeded pretty much routinely after that, though there were some problems with the hematocrit, a ratio reflecting the blood's capacity to carry oxygen. Also, they had a hard time coming off pump, meaning that Matthew's heart and lungs didn't want to function on their own, even with heavy doses of inotropes and vasodilators. But eventually he made it off, and Julie and Fackelmann and a couple of O.R. nurses were able to roll him to the PICU, Julie remarking after the baby was in the ICU and stable, "I think there was a little angel on his shoulder."

Thus were three sets of parents thrust into the agonal throes of watching over their babies following the Norwood procedure. Kenneth Mangan's surgery had gone well, though the lung damage he'd suffered in his first hours of life continued to plague him; he remained on the ventilator, but he was stable. Matthew Anderson was scrutinized continually, and within a day his scalp was covered with electrodes whose job it was to monitor brain waves. He, too, remained on the ventilator. Sommor Beeman, who had been moved over to the ward for her final days of recovery, remained the star hypoplast baby. By all appearances the range of the Norwood experience was accurately described by these three babies—though it would probably be more accurate to say that with HLHS, every case was completely unique. There was no comparison, and no way to predict the outcome.

Mee met with Tera and Anthony shortly after Matthew arrived in the PICU and gave them the less-than-hopeful news. The baby's aorta was smaller than they'd originally thought, and he'd had difficulty coming off pump. Mee told them he didn't know if Matthew would live—maybe not even an hour, he couldn't say. This was the worst news in all the days of what Tera described as "pure hell." They could only wait now, be with their baby. Anthony, dressed in jeans, and Tera, in sweats, hovered over Matthew. He lived an hour, then another. And another. His parents began to hope. Mee, Anthony said, had "basically told us to hope for the best and be prepared for the worst." He added, "But you never are." Tera, a woman with a clear, forthright manner, and cheeks that seemed perpetually red from tears, said that Matthew's remaining alive "builds your hope back up."

On the second day Matthew's doctors ordered him monitored to check for neurological insult. His responses were slow, and they wanted to rule out brain damage; medications could be the cause, and initial results were hopeful. Two days after his surgery Matthew seemed stable, and Tera, Anthony, and Tera's sister, Jamie, returned to Ronald McDonald House, where they had been staying. They were deeply exhausted that night but planned to get up early to be in Matthew's room the next morning when rounds began, to listen to the presentation and speak with Dr. Mee.

Tera woke from a deep sleep at two A.M., disoriented in the silent room. At first she thought it was morning already—hadn't that been the alarm clock? Full awareness arrived with a jolt: she must have been awakened by a ringing phone. Unable to make outside calls from her room, Tera ran downstairs to call the ICU. She reached Kim Teknipp. Yes, Kim had just called. Matthew had arrested; he was being resuscitated. "You need to get here," Kim said.

Tera bolted upstairs in a panic. She roused her husband and her sister, and the three of them (Tera dressed only in her pajamas) ran out to their car in the cold March night.

Charlie, the fellow on call, had opened Matthew's chest to administer direct massage and attempt to restart the heart. When Tera and Anthony and Jamie got to the clinic, they called from the fourth-floor lobby but were told to wait where they were, as resuscitation efforts were still going on. Five minutes, twenty, thirty. Finally Kim phoned Dr. Mee at home to ask if Demetrios Bourdakos, the on-call attending, could call the code, stop resuscitation, and let Matthew die. The Andersons and Jamie were still waiting in the lobby. Mee agreed. And then, spontaneously, Matthew's heart resumed its beating.

Kim went down to the lobby to get the family and take them back to Matthew's room—but only for a short time before his chest was rebandaged, left open. His heart had been shocked ten times.

Mee got to the hospital before seven A.M. and looked in on Matthew before meeting with his parents. He was blunt: they shouldn't expect their son to survive.

At eight A.M. the team crowded into the X-ray room to begin rounds, first examining the most recent films of Kenneth Mangan and Ashley Hohman, a tetralogy repair, a complex double-outlet right ventricle, and half a dozen others. This intro to rounds was typically brief and light in tone, the patients kept at a mental remove. But today the air was charged, the mood grim. Everyone knew what was happening two rooms down, if only by osmosis. The tension became overt when Bourdakos, looking gray, pushed into the room and caught Mee's eye.

"Roger," he said, "the means are in the twenties. I just pushed some epi."

Mee suggested, "Why don't you call the parents."

Bourdakos left to find Kim already on her way to Matthew's room with Tera, who was wiping her face, and Anthony, expressionless, eyes down. Bourdakos returned to the films room and told Mee, "We'll just follow the parents—if they want to hold him . . ."

"Absolutely," Mee said.

Rounds continued, as ever. Bourdakos slid a curtain along its track around the periphery of Matthew Anderson's room as members of his family began to fill up the lobby, becoming momentarily visible whenever the automatic doors opened. Dawn Blair, an ICU nurse, painted the baby's hands and feet blue and pressed them to paper, capturing handprints and footprints for his parents' memory box.

Tera, weeping, arched over her son. She had given Bourdakos permission to withdraw support—"He's been through enough," she felt. Bourdakos unhooked the IVs that had been delivering medications, then took out the breathing tube, gently picked Matthew up, and handed him to Tera. The baby boy never breathed on his own. "It's over," Tera said. "He's not suffering."

Rounding continued through the PICU for the next hour, and then the scrum as usual walked the hallway beside the offices to "the floor," or "the ward." Here, beside bed 2, Cindy and Doug Beeman were preparing for their trip home with their daughter, packed bags on the window seat, diaper bag and Sommor, in a carrier seat, on the bed. In the beginning Cindy had wanted to get to know everyone, be close to the other families, to share her story and feelings and listen to theirs. But as the possibility of home loomed larger, she had turned inward. "I've watched myself close off, which surprises me," she said. "I've cried enough. I'm done. I just want to go home.

"I don't think I could have another child," she continued, "for fear that something would happen like this again to my next child, and I don't want to bring another child into the world and make her go through this." She'd been told about the first hypoplasts Mee operated on here, who were still in apparent good health. "It's so nice to hear they're six or seven years old. We're afraid that [Sommor] will only make it to twenty and then she'll die on us, or that she'll only make it to ten."

As Doug checked their bags one last time and they got ready to vacate bed 2, Cindy said to him, "Maybe we can go out through the cafeteria. I know it's kind of germy, but I don't want to bring Sommor out past the family."

There were several ways to leave this floor, the least convenient of which was through the cafeteria. The easiest way was by using the elevators in the main lobby. But Cindy, like almost everyone on the floor, was aware of what was happening this morning, and she knew that Tera and Anthony's family had gathered in the lobby to be there for them. Her guilt at leaving the hospital with a recovered child was undiminished, but soon she'd be away from this place, she'd be home, her family intact.

This winter had been a difficult one in the ICU. While Mee was away at Christmastime—he took a month off each year to return to Australia, and this year to visit his ailing father in New Zealand—Jonathan had three complex cases that went badly, just a week before he was to depart to take over as chief surgeon at the children's hospital in Arkansas. Two patients died in the clinic, and a third would die later. "Jonathan's rattled," Marc Harrison said. He wanted to reassure Jonathan, "Nineteen out of twenty of those patients die—yours was one of the nineteen." Jonathan said it made him bitter—he went home to his wife and told her, "I don't want to operate anymore." The difficult winter had continued with the Anderson case.

The Monday after Matthew Anderson died seemed to be a routine day for Kenneth Mangan and his mother, Patricia. Patricia has six children at home, aged four to fourteen, and felt torn when her ten-year-old daughter begged her to attend her dance recital; she ended up going but was a wreck the whole time, thinking of Kenneth. She got to the clinic daily at eleven, and her husband, Brian, picked her up out front every night at seven. She did little more than hover over Kenneth, who remained sedated and on a ventilator after his emergency catheterization and his Norwood; his right lung was stiff and damaged. When the time came for her to leave each evening, after eight hours spent watching over an unconscious, mechanically ventilated baby, she could never believe it was seven already, and had to will herself out of the room. I'd gotten to know Patricia and Brian over their nearly three weeks here. All day long Patricia would watch Kenneth in his PICU room. She typically welcomed visitors. I once walked in and stood not a few feet away from her, on her immediate left. As she leaned on the warmer table over her son, I assumed she would notice me standing there. She whispered to her son, her nose just brushing his cheek. She kissed him. And she never saw me—she was focused to the exclusion of all else on her son, willing him forward.

On Monday, March 5, Brian Mangan put a TV dinner in the oven and told his fourteen-year-old daughter when to take it out. He had to go to the clinic to pick up Patricia. The snowstorm was so bad that he almost turned around at the end of their street to wait it out, but for some reason he decided to keep going. It was while he was on the highway that the call came from his wife. Don't pick me up outside, she instructed him; park the car and come up. When Brian got off the highway, the snow didn't stop, but seemed rather to disappear, as though it had never been snowing, as if he'd entered an out-of-time zone. The city streets were clear.

Patricia had been in Kenneth's room. Appachi was on call this night, dressed in slacks, dark shirt, tie, comfortable shoes. He wasn't doing anything, just watching the monitor as if thoughtlessly staring at some sports event. Nothing was wrong yet, and he said nothing to Patricia. But he didn't leave the room. The ICU was relatively calm; Appachi was not urgently needed elsewhere and so could just watch. The pressures dropped a little. He'd been caring for Kenneth now for nearly three weeks, along with everyone else, and none of them had ever seen a hypoplast have this kind of course, the intact septum followed by persistent pulmonary problems. Appachi folded his arms across his chest, set his feet wide apart, and stared at the monitor. Suddenly the baby's pressures plummeted, and his cardiac function fell. It all happened harrowingly fast, as the effects of sepsis often do. Kenneth's blood had become infected with a bacterium called pseudomonas, and he was going into septic shock. Appachi called for new lines to go into the baby's tiny vessels. As forces were marshaled, he had to ask Patricia to leave the ICU immediately. The baby needed large doses of fluids, inotropes, vasoconstrictors, and antibiotics right away; they might lose him, Appachi told Patricia. She was led out to the lobby to wait. She called Brian to tell him to park the car and come up. When a baby's blood becomes infected—here called gram-negative sepsis—the body's arteries go slack, resulting in low blood pressure and diminished heart function, and the organ tissue begins to starve, releasing harsh chemicals and lactic acid, which further impedes cardiovascular function. At this point, as Appachi's colleague Marc Harrison put it, "an unwillingness or inability to resuscitate aggressively may prove fatal. Unfortunately, even heroic measures are not always enough, and the child proceeds to deteriorate in a swift, terrifying death spiral."

Patricia waited in the quiet lobby. When the elevator doors slid open to reveal Brian, she began loudly weeping. As her husband stepped toward her, she said, "He's OK," and continued to cry.

Appachi had been able to get a line into their child quickly; he had delivered enough vasoconstrictors and fluids to clamp him down and maintain blood pressure; he had poured antibiotics into Kenneth, and they had gotten the bacteria in time. Maybe it was dumb luck that he'd been watching the monitor, or maybe some set of remembered patterns from previous experiences with kids who crashed had nagged at him enough to make him stop before the monitor. Whatever it was, the immediacy of his actions was likely responsible for Kenneth's recovery that night, and for his subsequent improvement. During rounds several days later Marc Harrison, looking at the film of Kenneth's improving lungs, said, "If this kid does well, don't write him up. I don't want any more like him."

Tamar Preminger, the interventional cardiologist who would have performed the procedure Prieto did if she had been on service that morning, said, "There aren't likely to be many more."

"He really gave us a run for the money," Marc said.

Patricia and Brian Mangan waited and prayed and on March 20 their baby boy was extubated and he breathed on his own for the first time. His parents brought the rest of the family in "to see his mouth," Patricia said, which had been covered with tape to hold the tube in all these weeks. To me she said with an easy smile, "Isn't it great? No tube in his mouth." Kenneth screamed when she kissed him, a real voice after a short lifetime of silence. "Are my lips cold?" she asked him. His cry had sounded scratchy to her, and she said, "Do you have a sore throat?" They were still in the ICU, but now, she said, "I can let myself think of bringing him home." It would be a couple more weeks still, but in early April Patricia and Brian would drive their son home.

In the world of congenital heart defects, families of children with hypoplastic left heart syndrome are a unified bunch. It is the only defect to which whole Web sites are devoted; there's also a book aimed specifically at parents of hypoplast babies, and a number of support groups to help them. At the clinic, social worker Nancy Petrov handles all hypoplast cases, and she organizes meetings several Saturdays each year for a group of parents whose children were born with HLHS. Occasionally Nancy arranges for guests to come speak to the parents. One weekend it was Maryanne Kichuk, who lectured and answered questions about transplantation (a possible necessity for single-ventricle hearts), and another Saturday Jessica Fletcher, the first baby born with the defect to be operated on in Cleveland and to have survived; she's nineteen years old and doing well. The surgeon who performed the operation, Carl Gill, made local headlines at the time, but as was the case at many centers during the early years of the Norwood, subsequent results were so dismal that the clinic for a time stopped offering the procedure. Every now and then Nancy persuades Mee to come in on a Saturday and speak to the group. Six or seven families typically attend these meetings, but when Mee speaks, the room fills up. ("They will tell you he walks on water," Nancy said to me of the parents' response to Mee. Gwen Ferchen's mother, Jeanne, regarded her grandchild Evan in PICU room 1 shortly after his modified Norwood procedure, and said, "They say Dr. Mee has the hands of God," and I could tell from the gravity with which she said this that she believed it.)

At the last meeting of hypoplast parents I attended, in February 2002, the Mangans brought along Kenneth, now one year old. Kenneth had big brown eyes and straight dark hair, and that day he was toddling unsteadily around the room, clutching for this grown-up's knee or that chair but usually landing with a *thud* on his bottom. He played with toys while the group conversed.

Patricia had been a wreck the previous August when she and Brian brought the baby in for his Glenn, the second of the three planned surgeries; she'd been short with Kim, the cardiac nurse handling preops that day, and Patricia told Mee this during their consultation.

"Keep your fists off her," he told Patricia. "She's a very nice lady."

"We're a little on edge," Patricia said.

"That's understandable. You'd be subhuman if you weren't."

"He looks so good now. We don't want him to be worse."

"He will be worse," Mee assured them.

And she was a wreck waiting for Kenneth's surgery to begin. Mee had trouble with his first surgery that morning; it took him more than an hour longer than he'd anticipated, so he had to go straight from that case into Kenneth's, breaking only long enough to scrub, regown, and glove. Once Mee was in the O.R., Patricia grew calm. In the early evening a nurse called down to the waiting room; Brian answered and heard her say, "Dr. Mee is very pleased."

In three or four years the Mangans planned to bring Kenneth back in for his third and final open-heart surgery.

One of the dads in this hypoplast group, John Radel, said, "I just *hate* coming through these doors, knowing I'm coming back." His daughter had been stalled in the ICU for weeks after her second operation because of a staph infection, but she was a lively "terrible" two right now, her parents' focus and delight. John's wife, Shelly, was also there; instead of working as a manufacturer's sales rep, as she had once done, she was now a stay-at-home mom, and active in the congenital-heart community. She'd just helped organize the second Congenital Heart Awareness Day here, a national effort the Sunday before Valentine's Day, intended to bring families together and raise awareness for patients whose illness was so little known and poorly understood. It had attracted hundreds of people this year, spearheaded by Mona Barmash, the founder of the Congenital Heart Information Network, an Internet site geared toward parents of children with congenital heart defects. Barmash's son, now a young adult, was born with tetralogy of Fallot and ultimately had three surgeries, the last one a terrible experience that prompted her to create her Web site. "I was a pissed-off parent," Barmash told me, "and I blamed myself because I didn't know enough to ask the right questions." She hoped, through her site, tchin.org, to help make parents become better consumers in a confusing medical world.

Also at the meeting was a woman who would soon give birth to a hypoplast baby; she wanted to learn as much as possible about what to expect. And, too, there was Beth Yannich, whose daughter Molly, now seven, was one of the first hypoplasts Mee operated on at the clinic. Today Molly is virtually unrestricted, despite her single-ventricle heart, though she has to have regular checkups with her cardiologist and must remain on Coumadin, the anticlotting drug that can turn a routine fall or head bump into something more dangerous (monthly blood tests are a requirement with this medication). Physically she is very active, but she is behind in school and has memory difficulties that impede her progress in math and reading. There will always be, for Beth and her husband, a fear of the unknown, a real uneasiness at the fact that no one knows what sort of life expectancy a successfully palliated hypoplast may have, but for now the blood tests and the school struggles—both of which will likely end or at least diminish with time—are their biggest problems.

In January 2002 I contacted Tera and Anthony Anderson and asked if I could come see them. They lived in a stationary trailer in Ravenna, Ohio, about thirty miles southeast of Cleveland. I arrived at dusk, and Tera answered the door wearing a loose-fitting shirt and sweatpants. Anthony appeared behind her; he'd been home from his job at Davey Tree Service for an hour or so. Their son Phillip played quietly in a room down the hall. The kitchen counters and table were clear, and the carpeted living room seemed to me immaculate, given the presence of a four-year-old boy with toys. A bookshelf holding books, pictures, and a television ran the width of the far wall. Tera sat cross-legged on one couch, I sat on an adjacent one, and Anthony sat in an easy chair on the opposite side of the room from Tera.

Tera said they were over the hard grief and could talk about Matthew now. She even seemed to welcome the chance to talk about him; some things about the whole experience had been troubling her for a long time. She told me she'd returned to the clinic a month or so after Matthew died to speak again with Dr. Mee. She wanted to clarify in her own mind what had happened. She had been traumatized at the time and hadn't clearly understood it all. It was a good meeting at first, with Mee explaining to her what had caused Matthew's death: what he had was like a heart attack, he said, the heart muscle not getting enough blood. But then Tera asked Mee a question, the answer to which was to upset her beyond words. "Would a transplant have worked?" And Mee replied yes, it might have. Tera's mind reeled. She didn't understand. No one had ever spoken to them about transplant, and yet they'd expressly said they wanted to do everything they could for Matthew.

"When he told me that transplant would have worked, that's when I lost some respect for Dr. Mee," she said. Tera tapped ash off her cigarette into the ashtray she held in her lap. "That made me mad." She said she'd been so angry she'd stood up and walked out of the room without saying a word.

I surmised aloud that many factors had likely been at work, that transplant wasn't an easy fix and was in fact a difficult solution on a number of counts, not least of which was finding a heart for a critically ill newborn, and that the situation probably hadn't been as simple as she thought. Transplants weren't the definitive answer people seemed to imagine they were. (I would later ask Mee why transplant had never been offered, and he'd tell me that because of organ scarcity, the protocol was always to attempt a Norwood first.

In Matthew's case, after that procedure, and perhaps even before it, he'd been too sick to be a transplant candidate.)

Tera and Anthony were under no illusions about bringing a baby with that condition home, and in one sense they felt extraordinary relief in knowing they'd done what they could, and in the fact that the baby had died quite quickly at the hospital, rather than at home after a protracted illness. Anthony was quiet, but when he spoke he was clear and thoughtful. "It would be hard to take him home when he was so sick," he remembered thinking. "If he's going to die, I hope he dies now." Occasionally Anthony slid a thumb and finger beneath his glasses to pinch the bridge of his nose, where tears formed. Tera likewise spoke between brief bouts of tears.

There was some good news, though: Tera was pregnant, due in two months. She and her husband were both happy and worried. They knew about the increased risk of heart defects, but Tera had been thoroughly examined, and her doctors had told her the baby appeared to be perfectly healthy. (And so he would be, and born on schedule, a boy they'd name Christopher.)

I had asked to visit Tera and Anthony at home for a couple of reasons, foremost among them to explain more fully what I was writing about and to ask permission to use their names, and Matthew's story, because the Norwood operation and the range of experience of parents whose children had HLHS mapped out one of the most visible and fundamental territories in the scary landscape of congenital heart disease. (All of the parents of whom I made this request eagerly agreed; invariably they said they hoped their stories might help others who found themselves in the same situation they'd been through.) Matthew seemed present in that living room and was still a powerful force in the couple's thoughts as they spoke about him. Two pictures of him hung on the walls, one with a poem about God's angels. I'd also wanted to visit with Tera and Anthony for another reason, out of a more personal and visceral curiosity, to try to get at least some small sense of the lasting impact of losing a baby in this way. It was Anthony who most clearly articulated that impact:

"You learn to value life," he said.

I noted that many parents I'd spoken with had said something similar, and I added, "You must hug Phillip with a special intensity."

Anthony leaned his head back in his chair, closed his eyes, and moaned in response—a low, almost primitive noise.

I looked at the photographs of Matthew and said, "He was a beautiful baby"—and it was true, he was, peaceful and dry and soft and innocent. Anthony said, "Perfect. Except for his heart."

We had spoken briefly of the costs. Had their insurance covered this, or had they had a difficult time of it? Insurance had covered almost all of it, Anthony said, bills totaling about $130,000. "But I'd have paid every dime myself," he added.

Obviously a year later they were moving forward in life—Anthony continued to work, their one child was thriving, Tera was pregnant—but I needed to know that there was some peace for parents whose babies died, some acceptable end to what they'd been through. Was there? Did the pain subside, did the door to this room close?

Anthony pinched the bridge of his nose and whispered, "*I think about him every day.*"

11. What's It All About, Anyway?

Long after that winter was over, after I'd left to hole up and try
to make sense of this world and the stories that never ended, I
spoke with Frank Moga. Every time I returned to the clinic for a
follow-up interview or to use the library there, it was always the
same routine: Mee heading into the operating room, diagrams of
bizarre pulmonary veins on his desk and empty packets of Nicorette
and an unfinished Pepsi, and Marc and Appachi and Steve and the
other docs in the PICU invariably shaking their heads when I
asked how it was going, saying, "Really busy"—which roughly
translated into "A lot of sick kids, not enough beds." By then
the new surgeon was adjusting to Mee's system, and Jonathan
Drummond-Webb was planted in Little Rock doing hundreds of
cases, and Frank was back at home in suburban Minneapolis, back
at the practice where he was beginning to do his first complex cases,
including his first Norwood. How did that happen, I asked, how
did it work? (Meaning, really, *What family let you do your first
Norwood on their child, and how did you handle it with them?*)

Frank explained that a hypoplast had been born in town with a
huge ascending aorta, an anomaly that significantly reduced the
risk. His partner, Dave Overman, had performed the surgery, with
Frank assisting. When it was done and the patient was stable in the
ICU, Frank had said to Dave, "Next hypoplast with a six-
millimeter aorta is mine."

A few weeks later, just such a heart had appeared. Dave talked it over with the referring cardiologist, and they decided Frank was ready. Frank discussed the case with the family, but he did not volunteer the information that this would be his first case. It was the grandmother who finally brought it up, Frank told me. She asked him, *Have you done many of these?*

Frank said, "No I haven't, but my partner has; we'll be doing this together and we're going to take very good care of this baby." And so they did. This was how the system replenished itself.

I worked my way through this story, tried to imagine an end— but what was an appropriate end to a world that never ended? There would forever be a woman cradling her baby on the other side of the operating-room doors, tearfully giving up her child to strangers, always that mom alone weeping into her hands on the other side of those doors. That was the true end to the story. And though she wept, though the pain was excruciating, like no other, the end was rarely so sad. Almost all of those moms, ninety-five out of a hundred of them, would hold their babies again after that most awful moment of handing them over to a stranger; they would feed their babies and change their diapers; often they would endure more surgeries, and they would invariably spend more time in cardiologists' offices. If the child had been transplanted, there would almost surely be more transplants, more agonized waiting for a heart, waiting that often took children into mortal danger before they could qualify for those very valuable, woefully limited organs—there would never be enough donor hearts to go around. But five or fewer of those one hundred moms would hover over their children—the hypoplasts, say—and give Demetrios the OK to withdraw the support and if they were lucky they would hold their babies while they were still warm, one could only pray, because the body became so cold, death was so quickly cold and ashen gray. One or two of those babies—maybe the baby born with pulmonary atresia and an intact ventricular septum—would do fine after surgery and move out of the ICU to the floor and then arrest there, among the recovering patients. He or she would have a right-ventricle-dependent coronary, and after a big feed, the blood would rush to the stomach and leave the delicately balanced coronary circulation, and the heart would arrest, and urgent attempts at resuscitation would be futile and the grandma would scream at Marc, who'd have called the code and would be actively trying to save the infant, "What are you doing to my baby!" and the baby would die and the mother would collapse in the middle of the hall in a paroxysm of anguish, and the next morning at eight A.M. rounds would begin as usual in the films room, and the O.R. would be readied for the day's cases.

When I spoke with Frank Moga, he said he had lost a patient. It was his first of what would certainly be more, as long as he continued to practice. I had seen Jonathan Drummond-Webb and Roger Mee lose cases, and Frank himself had once told me about the time he was scrubbed in on a case when he was a resident, and the surgeon made an error and the baby died in the operating room and then the surgeon told *Frank* to give the parents the news. "The most awful moment of my life," Frank had told me. Awful, tragic, horrific—but then he'd been a witness to and messenger of the horror, not the one responsible for it. Now, for the first time, *he* was responsible, it was his case. And the stress of it still made him feel like throwing up. It was a baby born with tetralogy, a repair that usually had a low mortality. As was customary at his center, the infant was shunted first, then scheduled for the repair before a year was out— but there was undiagnosed obstruction in the pulmonary artery. Had Frank known about this—had he asked to do an angiogram, he said—he could have planned for a procedure to fix it, but no one knew it was there, and the baby died three days after the operation. It had been two weeks now, and he was over the hard part, the first couple of days when he'd go home and wonder how he could possibly operate again. He'd never before run into a problem he couldn't get out of—he'd always been smart enough and lucky enough and ahead of the game enough, even in the worst situations. "And that scares the shit out of me," he said. "You can't do it perfectly every time, but that's the drive." He'd thought a lot about the hugeness of what he and everyone else in this specialty dealt with, the enormousness of the responsibility, had talked with me about it compellingly, artfully, and so convincingly that I knew *he* knew what he was talking about—and now here he was, telling me it was far bigger than even he had imagined it would be. For the first time he'd been questioning his choice, been asking himself, "What am I doing here?" This work was *hard*.

Jonathan Drummond-Webb had said much the same thing to me, about how periodically he would tell his wife he never wanted to operate again. I had spoken with Ed Bove when I visited his center at the University of Michigan, and he'd said he'd felt that way after he lost a series of Norwoods, a procedure he was renowned for. Another of the "celebrities" within the specialty, Frank Hanley, now at Stanford, said, "You have to be a certain kind of person to be able to look at yourself in the mirror—you're only human—and you know sometimes something you've done, or failed to do, has resulted in a death. It takes a certain kind of person to be able to keep working. We struggle with that all the time. . . . It's a black hole. You can find yourself over your head."

Jonathan said what got him back into the O.R. was that he eventually began thinking, "Maybe I can help someone." Frank didn't have to operate if he didn't want to—it wasn't the only way to make a living in the profession he'd chosen. He could move into a different specialty, could be a pediatric cardiologist, say, or a general pediatric surgeon. But what would happen if everyone, all the surgeons, acted on that thought? Sooner or later there'd be no one left to do the riskiest surgeries. Someone had to do them, because they could be done, and sometimes the surgeon who chose to do them would be average, and sometimes he would be excellent, and Frank wanted excellence, he wanted grace. Whether he had the physical skill and artful imagination and intuitive feel for the delicate physiology of these difficult repairs on complex organisms, these human babies, remained to be seen, but this was how a surgeon started out, and he was going to give it a roll because what was this life all about, anyway? "I'm just going to have to work harder to make my own luck," he said.

He uttered the words often: "What's it all about, anyway?" He asked the question in response to benign but inept care or political infighting that had nothing to do with patients. *It's about the patient,* was his unspoken answer. Farmer Frank who loved harvesting hearts—"It's so life-affirming," he had said with an electric grin. It was a good all-purpose query, that: What's it all about, anyway? Fackelmann had his own answer: *It's not about the money, dude. This is a kid here, this isn't a Yugo.* And Roger Mee had one, too. The best seemed to know what it was all about, but what I couldn't figure out was whether they were the best because they had known it most clearly from the beginning, and skill and excellence had followed, or if they'd been good at what they did first and *then* learned what it was all about. Frank hadn't known how life-affirming he'd find this work when he got into it. He had pursued an innate desire for big challenges and big results—the bigger, the better, because that was part of what his life was all about.

Roger Mee had been in cardiac surgery for more than thirty years, mainly working his ass off and not seeing his family. I knew by then that he was a decent man, as good as everyone had suggested, and I trusted his vantage point; he wasn't an intellectual wanker, but neither was he immune to reflection and comment.

One story he told me seemed particularly resonant because of his final off-the-cuff verdict on it. The case was his second transplant in Australia. A teenage boy needed a heart, but a stroke had left him slightly neurally impaired. At the time—this was in an era when many believed that Down's syndrome babies, for example (about half of whom are born with heart defects), ought not to be operated on—all potential transplant patients had to be evaluated by a team of psychologists and social workers, who decided whether they were worthy candidates for the unspeakably valuable commodity of a heart. The teenager failed the test, and his IQ was deemed insufficient to qualify him for transplant; he was thus sentenced to a death from heart failure with his parents by his side.

While in the boy's room one day, Mee noticed that he was playing what looked like a sophisticated electronic game. Mee thought that was curious—he shouldn't be able to do that if he was as badly off as the social workers said. He kept thinking about it and eventually sat down with the boy and had a talk with him. Mee sensed what the problem was and asked, "Are you pissed off you had to take that test?" And the boy, according to Mee, said, "You bet." Realizing that he'd purposely failed the test, Mee convinced him to retake it—it was only a formality, Mee told him; they'd make up their own minds. Mee then persuaded the social workers that the boy had had a "surprising recovery."

"We went ahead and transplanted him, and I'm still getting letters from his mother. . . . We were quite wrong to downgrade his value." The boy went on to become an enthusiastic ambassador for transplantation in the Australian media and, Mee said, was still doing well all these years later. "Good people," Mee noted, especially the kid's devoted mother.

"To me," he said, "the essence of life is love, and as far as I know, there's no test for that."

Of course, a child couldn't be denied care because he or she was unloved or came from a nasty family—the equation didn't work that way. But a parent's love for a child was all Mee needed to see before he'd try to help, if the parent asked, no matter how dismal the situation. When he first arrived at the Royal Children's Hospital in Melbourne, he said, "it was considered silly to operate on a Down's, but the parents *loved* these kids."

It really did come back to the children. *They* were what it was all about. Congenital heart surgery had come of age and could now be done with very low mortality, but it would take diligent work to maintain this standard, let alone improve it, and there was still, Mee said, "a lot of catching up to do at a lot of centers." He explained that fewer med students were going into cardiac surgery these days because the training took so long, and most young doctors would prefer to start earning more money right away, or maybe be with their families rather than wedded to the O.R. This was yet another reason he thought regionalism represented the future of the specialty, with fewer centers doing more cases (Ohio certainly didn't need *five* centers, for instance; this only diminished patient care, he believed). "I think the next step is regionalization," he said, "because then the resources can be concentrated." But even here he brought the focus back to the patient, as always. The child was what it was all about.

"When we look at an average family," he said, "we would expect them to put more than they can afford into their children. A community should do the same thing. I think that's the sign of a healthy community, when it's prepared to put more than it can afford into its children. I think that the most successful tribes historically have put a lot of effort into their children. It seems to be a good investment.

"All children ought to have universal health coverage. That would be the first step toward civilization. I don't give a damn about adults. They can make up their own minds. Children are innocent."

Out of the blue I'd gotten an e-mail from Frank asking if I'd seen a show on public television the night before, a drama called *The Innocents*, about the implosion of an English heart center after a period of unnecessarily high mortality—that is, after a number of babies died unnecessarily ("A powerful and absorbing true story of ill children, desperate parents and incompetent medical care based on the recent notorious Bristol Royal Infirmary scandal," according to promotional notes for the show; my wife couldn't bear to watch what the parents were going through and turned off the television with a horrified moan). Frank asked how my book was going (partly he was nervous, suspecting he'd talked too much about this closed guild), and I told him I was at what should be the end, but I didn't know what the end *was*—this world was endless. I always saw that woman sobbing into her hands, alone.

Frank, in his customary verselike e-mails, subsequently outlined a jokey sitcom-style ending for this book, but he also wrote one for which he couldn't maintain that lighthearted tone:

> go back to the clinic
> see the new faces, same as the others
> same as it ever was
> new sweaty parents
> same huge godawful inconceivable soulcrushing pressure
> looking for a healthy baby
> then go see H. or some other lucky family
> . . .
> [show] how they survived
> the stress the diamond making world changing stress
> and show that diamond man

Drew H. was whom this work was for, but he wasn't the diamond. He was one lucky boy. Lucky boy, and lucky parents who
loved their son with a depth that no one could fully know who
hadn't been through it. Judging from what I'd witnessed, there
might be no greater gift in the midst of tragedy than to know those
depths of love. Anthony Anderson knew those depths when he
wrapped his arms around his perfect son—that, too, was a kind of
gift, some sort of compensatory understanding forged by the soul-
crushing anguish of his loss. Even those parents who lost their children might sometimes know grace.

Drew H. was doing well. His parents, Angie and Bart, had their
moments of stress when Drew, forever on antirejection drugs that
disarmed his body's immune system, got sick in the middle of the
night or ran a high fever. But if their son now thrived, he was nevertheless not the diamond mentioned in Frank's e-mail. Rather, the
diamond was the surgeon, the technician, the ICU doc. One way or
another, the parents always left this world behind. They didn't feel
this pressure every day. They returned to a changed life, but one
that was their own; they found their own routines again. It was
Frank and Roger Mee and Fackelmann and Appachi and June the
cardiac nurse who had to live with the pressure of responsibility for
their actions, doing daily work that could quickly kill their patients. It was they who were hardened into what was surely rarefied
matter.

Drew H. was likely alive today because of Ashley Hohman. The
United Network for Organ Sharing had informed the clinic that
snowy night just before Thanksgiving of an available heart because
the heart was a match for *Ashley*, not for Drew. But it had turned
out to be just a little too big for Ashley, and in any event she was
septic at that point; even June Graney thought Ashley was on her
way out. Unlucky Ashley, Ashley Grace—the result of Tim and
Kelly Hohman's fourth and last in vitro attempt, whose heart had
begun to fail even before she was taken out of the womb in that
emergency C-section on September 20, who'd then been rushed to
the clinic, where echocardiography had diagnosed her strange le-

sion, a rudimentary ventricle composed not of standard-issue myocardium but rather of a spongy degenerative mutation of it. Jonathan had shunted Ashley to ensure that enough blood would get to her lungs. There was never any thought of *fixing* a heart like this; transplant was the Hohmans' only hope for their newly born child. And hope for it and pray for a transplant they did. On and off the list she went and came, and morning rounds were filled with debate as to whether or not she was even an acceptable candidate: her neural status was a question mark, her lungs were bad, the acites, fluid turning her belly into a watermelon.

Maryanne Kichuk didn't like it; she believed Ashley's heart was not her only problem.

During rounds one day Roger asked Jonathan, "Is she status seven?" Status seven meant still listed but currently ineligible.

"She's status one," Jonathan answered. "Internally she's not."

"What does that mean?"

"If one comes up, we can turn it down."

"I wonder if it's got out that we're not that keen on it," Mee said at the baby's bedside a week later, aware that Maryanne was ambivalent about listing her.

Frank said, "She's still status one."

"Internal status seven," Roger clarified.

"Right," Frank responded quickly, "but we're still aggressive and hopeful."

Kathy Weise asked, "What does internal seven mean?"

Frank replied, "It means we can do what we want. If there's sepsis, we can defer the heart."

In the PICU Maryanne told me that Ashley had never been a good candidate for transplant and that she wished "those two"—Jonathan and Roger—would stop giving Tim and Kelly false hope; a heart wasn't going to fix Ashley, and it could kill her. "Roger doesn't understand the issues," she said. "He's a great surgeon, but this is not his job. . . . It's a very precious commodity we're dealing with."

This was a salvage situation, Roger thought, and that was what transplants always were. The bottom line was that if she didn't get a heart, she was going to die. "I think we should give every kid a chance," he said. He felt Maryanne was too concerned with the surgical mortality figures for her transplant program.

Another week passed, and as rounds concluded and the staff filed out of Ashley's room, Mee said sternly to Maryanne, "Better use your charm and get her a heart."

"And deprive someone else of one?" Maryanne asked him.

Thanksgiving was especially difficult for Tim and Kelly, who spent the day in the ICU. They had been here for two months now. The day before, that new patient, Drew, the same size as Ashley, had gotten a new heart. He'd been here only a *day*, while Ashley'd been waiting eight weeks. Even Dawn, the nurse who spent the most time caring for their daughter, had openly wondered to the Hohmans about it that night—she couldn't understand, she said, why Ashley hadn't gotten that heart—and her remarks added fuel to their suspicions that the team wasn't going all out for them.

Another week passed. Kelly sat in the rocker doing word-search puzzles, fussed with Ashley's lines, cooed in her daughter's ear, whispered to her, and rocked some more. Jonathan drained more fluid from her belly. Kelly waited and watched and then left before midnight to return to Ronald McDonald House, praying as she turned off the light that there'd be a call in the middle of the night telling her that a heart had been found. Tim made the drive up after work most nights so he and Kelly could sit and watch their baby together. Kelly was relentless in her positive thinking: she'd been obsessively reading about how to care for a transplanted baby. She was already planning Ashley's diet, figuring out what foods she'd have to avoid because of the severe risk all transplants run of developing coronary artery disease. Kelly and Tim spoke almost casually about it, as if it were simply a matter of time. Tim didn't like the thought of having to give up sharing buttered popcorn with his daughter, but Kelly said she'd get him a hot-air popper. Eleven weeks and still no heart. Ashley's mother was becoming a well-known fixture in the PICU's daily life.

When Geoff Lane came in to check on Ashley, he asked her nurse, Dawn, "What did you do? These are the best sats I've seen in months."

"She likes the Christmas tree I brought her," Dawn said. A little

Christmas tree sat at the foot of the bed, along with a pile of stuffed animals and a swinging Santa figure programmed to sing "Jingle Bell Rock" at regular intervals.

PICU attending Steve Davis is sitting at the front desk at 6:25 P.M. when Cheryl Malek, the most senior of the cardiac nurses, beckons him away from the knot of ICU nurses. "Can I talk to you?" she asks.

"What's up?" he says, getting up from his chair.

Cheryl points to a white card in her hand, information about a heart donor. "I've paged Roger," she says. "I don't want to say anything till I know he can do it."

It's mid-December and Jonathan is taking a week off before Mee leaves for his vacation. That means Mee is the only surgeon; he's done three cases already today, with more on the way tomorrow and the day after that. And besides that, he's sick with a cold and the flu. His eyes are watery, and his skin is pale. He was eager to get home and into bed this frigid night, to try to shake his bug. Cheryl and Steve both know he's ill, and tired.

At six-thirty Cheryl is called to the phone. It's Mee, answering his page.

"Hi, Dr. Mee," she says. "Maryanne thinks they may have a heart for Ashley Hohman." Cheryl pauses to listen and then replies, "O-positive. The baby arrested Sunday, CPR was performed for twenty-three minutes. . . ." She carries on with the list of pertinent info—the baby's weight, existing heart defects: a small ASD, a PDA—and concludes, "I didn't want to say anything until I knew if you were going to be up to doing it."

Mee says, "We can't very well turn it down," and then he gives Cheryl instructions for the harvest. O.R. nurse Bob Cherpak strolls through the PICU on his way to the lockers, intending to change out of his scrubs and head home. Steve stops him and says, "Don't go anywhere." Cheryl calls Maryanne to tell her it's a go; after hanging up she asks Steve, "Do you want to tell her?"

"No, you tell her."

Kelly, who's been waiting nearly three months for this moment, is standing over her daughter when Steve and Cheryl enter the room. It's dimly lit in here, quiet and warm; Kelly's been adjusting some lines so they won't get tangled—just fussing, really. She doesn't say anything, just looks up at Cheryl.

"Maybe," Cheryl says.

" 'Maybe'?" Kelly looks disturbed.

"They may have a heart."

"Really?" Her breath leaves her. "Oh, my God," she says softly. She looks hard at Cheryl and Steve. "Oh, my God."

"Maybe," Cheryl says. There are always ifs.

Trying to gauge how serious this is, how iffy it may be, Kelly asks, "Should I call Tim?"

"Yes," Cheryl says.

Kelly smooths her daughter's hair, stares at her for a moment. "Santa came early, baby," she says. She looks again at Cheryl and Steve and says, "I just got the chills." Steve is quiet and grinning. Kelly doesn't want to leave any possibility out, so she checks: "You're not pulling my chain, are you?"

"We would never do that," Cheryl says.

Hard-edged Steve puts in, "Even I wouldn't do something like that."

Kelly asks when it'll happen—should Tim rush?

"It'll be a while," Steve tells her.

She says, "I'm excited, and I'm scared to death."

"Both are appropriate responses," Steve says.

Frank Moga is the fellow on call tonight. I find him surfing the Internet in the fellows' lounge, having just put a chest tube in a young post-Fontan patient. Cheryl has told him the news. He dials Fackelmann's home phone, and when Fackelmann answers and hears his voice, he says, "I can't fucking believe this"—it's been a long day, three cases, and he's dead tired. Frank says, "Get some sleep, I'll give you a call when we have a time."

Frank returns the phone to its cradle and tells me, "I'm nervous about this one." Ashley's still very sick, and as Maryanne has maintained all along, a new heart won't solve all her problems, yet the parents' hopes are all pinned on this night.

On the way back to the PICU I bump into Bruce Kasnik,

who's walking in the hall with his elder son, Kyle. When I first saw Bruce, a week ago, he looked gray and frightened; his newborn son Connor had just been rushed to the clinic with transposition, a big surprise to both parents. By now Connor's recovering quickly from his close call in the O.R., when Mac got into the duct, and the subsequent arterial-switch operation; he'll soon be going home. "We've talked about how this is the worst thing that's ever happened to us," Bruce says. "But there have been some great things about it. It's restored my faith a little in humanity."

(Mac, too, benefited. Now practicing in Tokyo, he would recall more than a year after the surgery, "That was one of the most frightening events in my professional life. Thanks to Dr. Mee, there was no catastrophic sequel, and that was the fortunate thing to me. That [event] gave me a great opportunity to become more careful, meticulous, and to be good.")

Tim Hohman is in Ashley's room when I return. "I didn't think the anxiety would go away, and it hasn't," he says. He walks to the bed where his daughter lies and kisses her grossly swollen belly.

Calculating how old Ashley will be when she gets her new heart, Kelly says, "She'll be three months old tomorrow."

Tim clarifies, having already done this: "Eighty-four days."

Who isn't superstitious in a pediatric intensive care unit, who doesn't believe in God, who doesn't cling to meaningful coincidence? Not anyone I've met. Eighty-four days is exactly twelve weeks. It's December 12, the twelfth day of the twelfth month, twelve days before Christmas Eve. Tim notes all the twelves, but he doesn't know that Ashley will also happen to be the twelfth and final pediatric transplant of this year. Kelly can't reach any of her family by phone and is feeling frustrated.

By nine P.M. Fackelmann is in the O.R. corridor, wearing a green clinic winter coat over his scrubs and gathering the equipment he'll need: Playmate cooler for the heart, sterile gauze, a retractor, ice, sterile bags, lactate solution for washing and storing the heart, Prolene sutures, stay stitches. Bob Cherpak is at the desk, and Fackelmann says, "Tell Roger this is a redo. They need to be in the O.R. at twelve-fifteen. We'll be back at one." George Thomas arrives with his portable perfusion case—he'll be the one who administers the cardioplegia to stop the donor heart. Frank arrives with his arctic wear jacket over his scrubs, his plastic surgeon's clogs. They meet Earl Hazelton, a nurse from over in adult who'll do any coordinating that's necessary while the others work, and together they all head downstairs to the police escort that will take them to the waiting jet.

It'll be a quick flight to a mid-Atlantic state. There are deli sandwiches laid out on one of the jet's seats, and Fackelmann, who isn't feeling well, takes a rear seat facing forward so he can sleep. An ambulance is waiting at the airport when the jet lands. "How fast do you want to get there?" the driver asks, and Frank, recalling the last ambulance ride on a snowbound highway, says, "Alive, just get us there alive." A man named Rex greets the group at a back entrance and leads them through the hospital corridors to the O.R. Fackelmann now becomes upbeat, charged by the energy of his work (the energy always kicks in when he gets to the donor hospital, he says); he's been here before, he realizes, and notes that the hallways have been repainted. The baby isn't ready when the team enters the O.R.; Fackelmann can't believe they don't have lines in yet, don't know the pressures ("Mickey Mouse stuff," he says under his breath). Then, finally, they roll the baby in, and Fackelmann and Frank tenderly transfer him from the warmer bed to the O.R. table. Frank reads all the documents that have been prepared for him—certificates pronouncing brain death, organ-donor papers—and signs them. Fackelmann is clearly unimpressed by the way things are being run in this O.R., but he and Frank are all friendliness and grace, grateful guests. Soon they're scrubbed and opening, and when Frank has a clear view of the heart, Earl asks, "How's it look, Frank?"

"The anterior RV is a little bruised—Mike and I think it's from the CPR. The LV looks snappy." He's able to lift it up and examine the apex and the posterior side of the heart. "Hey, Earl, you can give them a call and tell them it's a go."

"What's the cross-clamp time?" he asks.

"Tell them twenty minutes."

"Tell them forty," Fackelmann calls to Earl, then turns to

Frank, across the table from him, and says, "OK, Frank? I don't want them to hurry."

"Tell them thirty. Sold!"

Fackelmann likes to teach, and so he offers a brief anatomy lesson to the scrub nurses, who don't see a lot of hearts. "See this IVC?" he asks, pointing out the inferior vena cava. "Usually we have to fight over this with the liver guys." George runs the plege, and they stop the heart. Fackelmann says, "Nice, quick arrest. See how the blood rushed out of the heart?" Then, a few minutes later, "You can turn off the lungs."

They remove the heart. Fackelmann stores it in sterile solution in ice in the big deli cup in the Playmate. A nurse tells Frank, "We'll close if you'd like."

"Yeah?" Frank says.

"They're scrubbing now."

"That would be great," Frank says. "Kinda sad business."

Fackelmann is thanking the O.R. staff, conveying his appreciation. A nurse says, "We'll be glad to know your little girl is getting it."

Fackelmann pauses over this baby who stopped living before his eyes. He is not draped, a naked infant lying flat on the table, his chest open and empty. The boy is just seven days old, with fine blond hair, cherubic cheeks, and pure, blemishless skin; he stopped breathing, for unknown reasons, two days ago. Fackelmann is quiet for a moment, regarding this child, and then he says, "Hey, what a beautiful kid, eh, Frank?"

Frank stops to look at the child. Then he turns away.

Frank is responsible for the heart while the other members of the team prepare to leave—gathering equipment, using the bathroom— and then they all duck back into the ambulance for the ride to the airport, then back into the jet, now dark and cold as a freezer, for the return flight. It's a smooth trip, and everyone sleeps as the wing lights flash in the frigid night. On the ground in Cleveland, nearing the clinic, Fackelmann grows sentimental and says to Frank, "That's probably our last transport together." The transport team walks purposefully through the doors of the clinic's Children's Hospital entrance shortly before two A.M., and into an elevator being held by security, which ascends to the fourth floor and opens on a lobby filled with Ashley Hohman's family. The relatives turn when they hear the elevator doors open, and they smile with excitement at Frank, who's carrying the red Playmate with the heart, and at the others, big, bright, happy smiles. Frank, when he's past them, mutters, "That's always the creepiest part of the night."

Mee is in his O.R. Ashley's chest is open before him. Fackelmann scrubs in to assist. When Frank enters, Mee asks, "Good heart?"

"I left the ductus, and the PA is short. I was going to scrub in and divide these now"—ligate the duct before Mee connects the heart. The heart—the focus of an errand freighted with a strange combination of grief and life and loss and hope—will start almost immediately, jump straight into sinus rhythm, as if it knew.

Mee will be out of the O.R. by four-thirty, leaving Fackelmann to close. The next case will begin in a few hours.

SOURCES AND ACKNOWLEDGMENTS

National data concerning congenital heart defects and their repair are hard to come by because there is no national reporting system or database. The numbers that are available tend to be published by single centers, but some studies offer data from numerous centers on thousands of patients. My information on the incidence of various lesions comes from a few sources.

One of the strongest sources on incidence remains "The Report of the New England Regional Infant Cardiac Program" by D. C. Fyler (*Pediatrics* 65 [1980]: 376–461). Reports from the University Hospitals Consortium, a group of more than sixty hospitals, have been helpful with data on specific lesions (see H. P. Gutgesell, T. A. Massaro, and I. L. Kron, "The Arterial Switch Operation for Transposition of the Great Arteries in a Consortium of University Hospitals," *American Journal of Cardiology* 74 [1994]: 959–60; H. P. Gutgesell and T. A. Massaro, "Management of Hypoplastic Left Heart Syndrome in a Consortium of University Hospitals," *American Journal of Cardiology* 76 [1995]: 809–11; and H. P. Gutgesell and J. Gibson, "Management of Hypoplastic Left Heart Syndrome in the 1990s," *American Journal of Cardiology* 89 [2002]: 842–46.) I have also relied on data provided by the American Heart Association's *2002 Heart and Stroke Statistical Update* regarding the number of open-heart surgeries and catheterizations, as well as information on heart disease generally. Dr. Richard Van Praagh drew on fifty years of data from his institution, Children's in Boston, to provide or confirm data at my request. Another excellent source I relied on for information concerning specific lesions and their repair is *Cardiac Surgery of the Neonate and Infant* by Aldo Castañeda, M.D., Ph.D., Richard Jonas, M.D., John E. Mayer, Jr., M.D., and Frank L. Hanley, M.D. (Philadelphia: W. B. Saunders Company, 1994).

For historical information about the development of cardiac surgery, I relied most heavily on *King of Hearts: The True Story of the Maverick Who Pioneered Open-Heart Surgery* by G. Wayne Miller (New York: Times Books, 2000), as noted in chapter 2; on *Saving the Heart* by Stephen Klaidman (Oxford: Oxford University Press, 2000); on a paper titled "50 Years of Pediatric Cardiology and Cardiac Surgery: A Pathologist's View" by Dr. Van Praagh, presented at Frontiers in Diagnosis and Management of Congenital Heart Disease, Newport, Rhode Island, October 25–27, 2001; and on *Landmarks in Cardiac Surgery* by Stephen Westaby, M.D., with Cecil Bosher (Oxford: ISIS Medical Media, 1997).

For information on heart embryology, I relied primarily on descriptions from *Langman's Medical Embryology*, 6th ed., by Jan Langman, ed. T. W. Sadler, Ph.D. (Baltimore: Williams & Wilkins, 1990).

Data from Boston's Children's Hospital, noted in chapter 6 are from Aldo R. Castañeda, "The Neonate with Critical Congenital Heart Disease: Repair—a Surgical Challenge," *Journal of Thoracic and Cardiovascular Surgery* 89 (1989): 869–75.

For information in chapter 9 on the events surrounding the Bristol Royal Infirmary congenital heart center, I relied on reports from the BBC.

The data by Kathy Jenkins and her colleagues regarding the volume-outcome relationship addressed in chapter 9 are from the paper "In-Hospital Mortality for Surgical Repair of Congenital Heart Defects: Preliminary Observations of Variation by Hospital Caseload" (*Pediatrics* 95, no. 3 [1995]: 323–30). She also directed me to the previously mentioned UHC reports by Gutgesell, along with the following studies:

Chang, R.-K. R., and T. Klitzner. "Can Regionalization Decrease the Number of Deaths for Children Who Undergo Cardiac Surgery? A Theoretical Analysis." *Pediatrics* 109, no. 2 (2002): 173–81.

de Leval, M. C., et al. "Human Factors and Cardiac Surgery: A Multicenter Study." *Journal of Thoracic and Cardiovascular Surgery* 119, no. 4 (2000): 661–72.

Hannan, E. L., et al. "Pediatric Cardiac Surgery: The Effect of Hospital and Surgeon Volume on In-Hospital Mortality." *Pediatrics* 101, no. 6 (1998): 963–69.

Hewitt, M., for the Committee on Quality of Heath Care in America and the National Cancer Policy Board. *Interpreting the Volume-Outcome Relationship in the Context of Health Care Quality, Workshop Summary* (an Institute of Medicine report available at www.nap.edu; examines the subject as it affects complex cancer procedures).

Scott, W. A., and D. E. Fixler. "Effect of Center Volume on Outcome of Ventricular Septal Defect Closure and Arterial Switch Operation." *American Journal of Cardiology* 88 (2001): 1259–63.

Sollano, J. A., et al. "Volume-Outcome Relationships in Cardiovascular Surgery." *Journal of Thoracic and Cardiovascular Surgery* 117, no. 3 (1999): 419–30.

Stark, J., et al. "Mortality Rates After Surgery for Congenital Heart Defects." *Lancet* 355 (2000): 10004–7.

"I am not aware of any studies refuting volume-outcome relationship," Jenkins notes, "although many people have noted exceptions."

In chapter 10, for information (in addition to that from sources already noted) on hypoplastic left heart syndrome, I have used numbers and information from the following reports:

Brackley, J. K., et al. "Outcome After Prenatal Diagnosis of Hypoplastic Left-Heart Syndrome: A Case Series." *Lancet* 356 (2000): 1143–47.

Doty, D. B., et al. "Hypoplastic Left Heart Syndrome: Successful Palliation with a New Operation." *Journal of Thoracic and Cardiovascular Surgery* 80 (1980): 148–52.

Litwin, S. B., R. Van Praagh, and W. F. Bernhard. "A Palliative Operation for Certain Infants with Aortic Arch Interruption." *Annals of Thoracic Surgery* 14 (1972): 369–75.

Mahle, W. T., et al. "Survival After Reconstructive Surgery for Hypoplastic Left Heart Syndrome: A 15-Year Experience from a Single Institution." *Circulation,* Suppl. 3 (2002): 136–41.

Norwood, W. I., J. K. Kirklin, and S. P. Sanders. "Hypoplastic Left Heart Syndrome: Experience with Palliative Surgery." *American Journal of Cardiology* 45 (1980): 87–91.

Norwood, W. I., Peter Lang, and Dolly Hansen. "Physiologic Repair of Aortic Atresia–Hypoplastic Left Heart Syndrome." *New England Journal of Medicine* 308 (1983): 23–26.

Ohye, Richard G., M.D., and Edward L. Bove, M.D. "Current Status of Operations for Hypoplastic Left Heart Syndrome." *American College of Cardiology Current Journal Review,* Jan./Feb. 2001: 84–88.

Van Praagh, R., et al. "Interrupted Aortic Arch: Surgical Treatment." *American Journal of Cardiology* 27 (1971): 200–211.

I would like to thank all the clinic physicians, who were in almost every instance generous to a fault in responding to my questions for a year and a half. Roger Mee, of course, gave me countless hours of his time, but others such as Marc Harrison and Adel Younoszai were also generous and helpful beyond what is revealed in the text. I would also like to thank Richard Jonas and Ed

Bove and his partner, Rick Ohye, for their generosity during my visits to their centers in Boston and Michigan. Richard Van Praagh was particularly helpful in giving me details and perspective in historical matters. Deborah Gilchrist worked effectively on my behalf while I was at the clinic and in helping answer myriad questions after I left; I don't know how I'd have managed without her help.

The author would also like to thank Elizabeth Kaplan and Ray Roberts, Bruce Giffords, Dorothy Straight, Steve Bickston, Katie and Paul Buttenwieser, Richard Ruhlman, and Donna Turner Ruhlman.

RESOURCES FOR PARENTS

This book isn't meant to be a guide for parents who find themselves thrust into the world of congenital heart disease, but I can point parents to certain resources and guides, as well as offer the advice of a layperson who has had a limited but intimate glimpse of it. One of your first resources might be an organization called the Congenital Heart Information Network, at www.tchin.org, which to my mind is a helpful starting point. While the Web site doesn't recommend actual centers or doctors, its links, book reviews, support groups, Q-and-A topics, information on specific lesions, personal stories, and guide to centers around the country provide a broad range of important information and services to help parents make good decisions in a confusing medical world.

As a general rule, if you use the Internet to obtain information regarding your child, do so with skepticism. Quality and accuracy of information is not regulated. "Many people use the Internet to find information specific to their child's condition," says Mona Barmash, founder and president of the Congenital Heart Information Network. "Before you begin your search for information, please understand that the resources you find may not accurately reflect your child's specific situation, especially if your child has more than one defect or condition." She offers the following sites as good starting points.

The Congenital Heart Information Network
www.tchin.org

An international organization that provides reliable information, support services, and resources to families of children with congenital heart defects and acquired heart disease and to adults with congenital heart defects, as well as the professionals who work with them

The Care Notebook
http://mchneighborhood.ichp.edu/CARENotebook/care-notebook.htm

A downloadable file that families can use to help organize information about a child's health and care, making it easier to share key information with the child's care team

The Heart Center Encyclopedia
www.cincinnatichildrens.org/Health_Topics/heart-encyclopedia/default.htm

Definitions and descriptions of congenital heart disease

North American Society of Pacing and Electrophysiology
www.naspe.org

Arrhythmia information, including a glossary and explanations of diagnostic procedures and treatments

MedLine Plus: Congenital Heart Disease
www.nlm.nih.gov/medlineplus/congenitalheartdisease.html

Information from the National Library of Medicine

After a lengthy immersion and long observation in one heart center, and having interviewed scores of doctors at that center and throughout the country, I've developed certain convictions that, taken with the caveat that I am a layperson and not an expert, might be helpful to a parent trying to do what's best for a child with a serious congenital heart defect. Given the complexity and difficulty of this medical specialty, numerous physicians noted that, as chief cardiologist Tamar Preminger put it, "Selecting the best doctor (and center) can significantly impact outcome. This is not the time to be timid; it is probably one of the most, if not *the* most, important decisions parents will be making in their child's life."

Here is what I've learned about how to make decisions and questions to ask if your child needs treatment for congenital heart disease.

Know that you have time to make decisions. Only rarely these days do defects require emergency surgery. Take your time. Ask questions. Ask for second opinions. Ask for the names of the experts in the field. Many parents, I am told, feel nervous about requesting a second opinion and fear that they will be ostracized if they question their cardiologist. Never be afraid of what your cardiologist will think if you ask questions. Any parent who feels, or is made to feel, uncomfortable at such a request is not at the right place. The best physicians will readily refer patients to other centers if asked. Much is out of your control, but what is in your control is the ability to get your child to the best doctor and center for his or her condition.

All medical records pertaining to your child, including angiograms and echocardiograms, are your property. Though some hospitals may charge an administrative fee for sending them to you, they cannot withhold them from you. When you are seeking a second opinion, most centers will review these records and offer an opinion on your child's situation as a courtesy. The most expedient way to do this is with the help of your cardiologist, who will forward the records directly to another center.

Centers can vary considerably in how successful they are. Success in the specialty of pediatric heart surgery has been correlated with number of cases performed. That a center is big doesn't necessarily mean it's good, but the bigger centers performing heart surgery on babies and children tend to have better results. When considering treatment at a center, you should ask how many cases are performed there annually and what the mortality is. The biggest centers perform 750 or more cases, moderately sized centers 250 to 500; centers performing fewer than 100 are considered small. Traveling to a center that specializes in a high volume of complex cases can be worth all the inconvenience and expense it may entail, and at the major institutions, social service and financial personnel can often assist with financial aid for travel.

Do know that by traveling to a different center for a surgical procedure, you are not abandoning your home institution and doctors; you will be referred back to your home institution for follow-up care. This is a small field, and your child can have many experts involved in his or her care without an extended commitment to long-distance travel.

If your child requires a complex or high-risk repair, ask your cardiologist which are the five best centers for your child's lesion. When you are referred to a specific surgeon, ask how many such repairs the surgeon has performed and what his or her mortality is (as opposed to a generally accepted mortality for that lesion, or that surgeon's overall mortality). Ask if this surgeon operates on adult patients with acquired heart disease in addition to congenital heart disease (again, a surgeon who performs a greater number of complex congenital cases tends to have a higher overall success rate for such defects). Compare this data to the average mortality at all centers, if the data are available. Ask your cardiologist for help in obtaining information. In some states, such as New York and Ohio, this information is available to the public through the state Department of Health.

Other questions to factor into your decision: Are the center's perfusion and anesthesiology teams devoted exclusively to neonatal and congenital heart surgery? (Again, the quality of congenital repair can be diluted at those centers serving both groups: children with congenital heart disease and adults with acquired heart disease.) Does the center have an ICU devoted only to pediatric cardiac patients? Are the intensivists, ICU nurses, and respiratory therapists trained specifically for cardiac patients? Is there a senior, or attending, ICU physician at the hospital at all times?

"For those new to the experience of raising a child with CHD," says Barmash, "keeping a file with current medical records, including test results and reports from procedures, can be extremely helpful. It is always advisable to take your child's most current medical reports with you when you travel, in case of an emergency or illness. Having things on file will also make it much easier to get a second opinion, as you can forward pertinent information to the consulting physician quickly and without having to sign releases. Ask to be cc'd on all physicians' correspondence regarding your child's case, so that you are receiving all new information on a consistent basis.

"Your child's medical records can also be a wonderful teaching resource. I typically recommend that parents make a copy of the reports they receive, and then highlight the phrases or words that they do not understand. Use these records to formulate questions to ask your child's health care team. Once you begin to understand the medical language and the concepts, you will be better able to understand any changes in your child's status and can formulate more detailed questions that will help you in decision-making processes.

"Many people find it helpful to have their physician give them a diagram of their child's heart. A detailed diagram, along with a good explanation by a member of your medical team, may be the best way for you to understand how your child's heart works and what may need to be done to help the heart work better."

As a rule, physicians and heart centers have your child's best interests in view; but always keep in mind that you, the parent, are the ultimate advocate for your child.